Radio Electronics and Engineering

Radio Electronics and Engineering

Edited by
Oliver Dunbar

Larsen & Keller
www.larsen-keller.com

Radio Electronics and Engineering
Edited by Oliver Dunbar
ISBN: 978-1-63549-690-1 (Hardback)

© 2018 Larsen & Keller

⊟ Larsen & Keller

Published by Larsen and Keller Education,
5 Penn Plaza,
19th Floor,
New York, NY 10001, USA

Cataloging-in-Publication Data

Radio electronics and engineering / edited by Oliver Dunbar.
     p. cm.
Includes bibliographical references and index.
ISBN 978-1-63549-690-1
1. Radio. 2. Electronics. 3. Engineering. I. Dunbar, Oliver.
TK6550 .R33 2018
621.384--dc23

For more information regarding Larsen and Keller Education and its products, please visit the publisher's website www.larsen-keller.com

Table of Contents

Preface

As a part of electronic engineering, radio electronics refers to the study of devices that operate between 3kHz to 300GHz radio frequency spectrum. This subject is used in almost every object that transmits or receives radio waves, like mobile phones, satellites, radios, Wi-Fi, etc. The main focus of radio electronics is to control and manage coverage and to using signals between transmission systems. This book elucidates the concepts and innovative models around prospective developments with respect to radio electronics and engineering. The topics covered in this extensive textbook deal with the core aspects of the area. Those with an interest in the field of radio electronics and radio engineering would find it helpful.

Given below is the chapter wise description of the book:

Chapter 1- Radio is the instrument used to convey information by the medium of sound. A radio consists of a transmitter, a receiver, antenna, etc. Software-defined radio and cognitive radio are some examples of radio communication systems. This is an introductory chapter which will introduce briefly all the significant aspects of radio and radio frequency engineering.

Chapter 2- Radio frequency can be measured in units called hertz. Radio frequency engineering is a field of study of electrical engineering which studies devices that are used to control radio frequency spectrum. This chapter is an overview of the subject matter incorporating all the major aspects of radio frequency.

Chapter 3- Radio receiver, radio transmitter design, radio repeater, absorption wavemeter, crystal radio and lecher lines are the various kinds of radio electronics. Radio receivers are devices used to convert radio waves into information which can be carried forward. This chapter has been carefully written to provide an easy understanding of the various radio electronics.

Chapter 4- Antennas are devices used to convert electrical energy into radio waves. Antennas are mainly used in radio receivers and radio transmitters. Some of the types of radio antennas are dipole antenna, adcock antenna, conformal antenna, dielectric resonator antenna and cage aerial. Radio antenna is best understood in confluence with the major topics listed in the following chapter.

Indeed, my job was extremely crucial and challenging as I had to ensure that every chapter is informative and structured in a student-friendly manner. I am thankful for the support provided by my family and colleagues during the completion of this book.

Editor

An Introduction to Radio

Radio is the instrument used to convey information by the medium of sound. A radio consists of a transmitter, a receiver, antenna, etc. Software-defined radio and cognitive radio are some examples of radio communication systems. This is an introductory chapter which will introduce briefly all the significant aspects of radio and radio frequency engineering.

Radio

The Alexandra Palace, here: mast of the broadcasting station

Radio is the technology of using radio waves to carry information, such as sound, by systematically modulating properties of electromagnetic energy waves transmitted through space, such as their amplitude, frequency, phase, or pulse width. When radio waves strike an electrical conductor, the oscillating fields induce an alternating current in the conductor. The information in the waves can be extracted and transformed back into its original form.

Radio systems need a transmitter to modulate (change) some property of the energy produced to impress a signal on it, for example using amplitude modulation or angle modulation (which can be frequency modulation or phase modulation). Radio systems also need an antenna to convert electric currents into radio waves, and radio waves into an electric current. An antenna can be used for both transmitting and receiving. The electrical resonance of tuned circuits in radios allow individual stations to be selected. The electromagnetic wave is intercepted by a tuned receiving antenna. A radio receiver receives its input from an antenna and converts it into a form that is usable for the consumer, such as sound, pictures, digital data, measurement values, navigational positions, etc. Radio frequencies occupy the range from a 3 kHz to 300 GHz, although commercially important uses of radio use only a small part of this spectrum.

A radio communication system sends signals by radio. The radio equipment involved in communication systems includes a transmitter and a receiver, each having an antenna and appropriate terminal equipment such as a microphone at the transmitter and a loudspeaker at the receiver in the case of a voice-communication system.

Classic radio receiver dial

Etymology

The term "radio" is derived from the Latin word "radius", meaning "spoke of a wheel, beam of light, ray". It was first applied to communications in 1881 when, at the suggestion of French scientist Ernest Mercadier, Alexander Graham Bell adopted "radiophone" (meaning "radiated sound") as an alternate name for his photophone optical transmission system. However, this invention would not be widely adopted.

Following Heinrich Hertz's establishment of the existence of electromagnetic radiation in the late 1880s, a variety of terms were initially used for the phenomenon, with early descriptions of the radiation itself including "Hertzian waves", "electric waves", and "ether waves", while phrases describing its use in communications included "spark telegraphy", "space telegraphy", "aerography" and, eventually and most commonly, "wireless telegraphy". However, "wireless" included a broad variety of related electronic technologies, including electrostatic induction, electromagnetic induction and aquatic and earth conduction, so there was a need for a more precise term referring exclusively to electromagnetic radiation.

The first use of *radio-* in conjunction with electromagnetic radiation appears to have been by French physicist Édouard Branly, who in 1890 developed a version of a coherer receiver he called a *radio-conducteur*. The radio- prefix was later used to form additional descriptive compound and hyphenated words, especially in Europe, for example, in early 1898 the British publication *The Practical Engineer* included a reference to "the radiotelegraph" and "radiotelegraphy", while the French text of both the 1903 and 1906 Berlin Radiotelegraphic Conventions includes the phrases *radiotélégraphique* and *radiotélégrammes*.

The use of "radio" as a standalone word dates back to at least December 30, 1904, when instructions issued by the British Post Office for transmitting telegrams specified that "The word 'Radio'... is sent in the Service Instructions". This practice was universally adopted, and the word "radio" introduced internationally, by the 1906 Berlin Radiotelegraphic Convention, which included a Service Regulation specifying that "Radiotelegrams shall show in the preamble that the service is 'Radio'".

The switch to "radio" in place of "wireless" took place slowly and unevenly in the English-speaking world. Lee de Forest helped popularize the new word in the United States—in early 1907 he founded the DeForest Radio Telephone Company, and his letter in the June 22, 1907 *Electrical World* about the need for legal restrictions warned that "Radio chaos will certainly be the result until such stringent regulation is enforced". The United States Navy would also play a role. Although its translation of the 1906 Berlin Convention used the terms "wireless telegraph" and "wireless telegram", by 1912 it began to promote the use of "radio" instead. The term started to become preferred by the general public in the 1920s with the introduction of broadcasting. ("Broadcasting" is based upon an agricultural term meaning roughly "scattering seeds widely".) British Commonwealth countries continued to commonly use the term "wireless" until the mid-20th century, though the magazine of the British Broadcasting Corporation in the UK has been called Radio Times since its founding in the early 1920s.

In recent years the more general term "wireless" has gained renewed popularity, even for devices using electromagnetic radiation, through the rapid growth of short-range computer networking, e.g., Wireless Local Area Network (WLAN), Wi-Fi, and Bluetooth, as well as mobile telephony, e.g., GSM and UMTS cell phones. Today, the term "radio" specifies the transceiver device or chip, whereas "wireless" refers to the lack of physical connections; thus equipment employs embedded *radio* transceivers, but operates as *wireless* devices over *wireless* sensor networks.

Processes

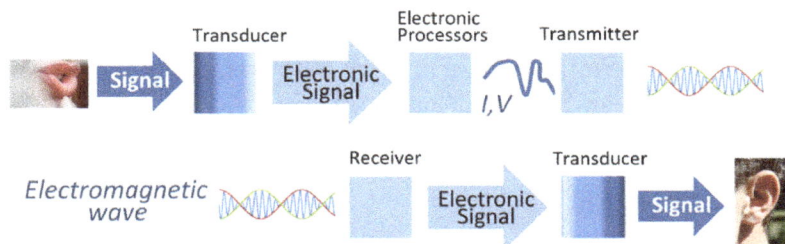

Radio communication. Information such as sound is converted by a transducer such as a microphone to an electrical signal, which modulates a radio wave sent from a transmitter. A receiver intercepts the radio wave and extracts the information-bearing electronic signal, which is converted back using another transducer such as a speaker.

Radio systems used for communication have the following elements. With more than 100 years of development, each process is implemented by a wide range of methods, specialised for different communications purposes.

Transmitter and Modulation

Each system contains a transmitter, This consists of a source of electrical energy, producing alternating current of a desired frequency of oscillation. The transmitter contains a system to modulate (change) some property of the energy produced to impress a signal on it. This modulation might be as simple as turning the energy on and off, or altering more subtle properties such as amplitude, frequency, phase, or combinations of these properties. The transmitter sends the modulated electrical energy to a tuned resonant antenna; this structure converts the rapidly changing alternating current into an electromagnetic wave that can move through free space (sometimes with a particular polarization).

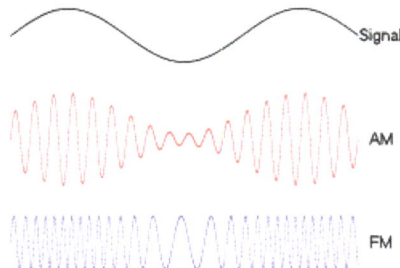

An audio signal (top) may be carried by an AM or FM radio wave.

Amplitude modulation of a carrier wave works by varying the strength of the transmitted signal in proportion to the information being sent. For example, changes in the signal strength can be used to reflect the sounds to be reproduced by a speaker, or to specify the light intensity of television pixels. It was the method used for the first audio radio transmissions, and remains in use today. "AM" is often used to refer to the medium wave broadcast band, but it is used in various radiotelephone services such as the Citizen Band, amateur radio and especially in aviation, due to its ability to be received under very weak signal conditions and its immunity to capture effect, allowing more than one signal to be heard simultaneously.

Frequency modulation varies the frequency of the carrier. The instantaneous frequency of the carrier is directly proportional to the instantaneous value of the input signal. FM has the "capture effect" whereby a receiver only receives the strongest signal, even when others are present. Digital data can be sent by shifting the carrier's frequency among a set of discrete values, a technique known as frequency-shift keying. FM is commonly used at Very high frequency (VHF) radio frequencies for high-fidelity broadcasts of music and speech. Analog TV sound is also broadcast using FM.

Angle modulation alters the instantaneous phase of the carrier wave to transmit a signal. It may be either FM or phase modulation (PM).

Antenna

Rooftop television antennas. Yagi-Uda antennas like these six are widely used at VHF and UHF frequencies.

An *antenna* (or *aerial*) is an electrical device which converts electric currents into radio waves, and vice versa. It is usually used with a radio transmitter or radio receiver. In transmission, a radio transmitter supplies an electric current oscillating at radio frequency (i.e. high frequency AC) to the antenna's terminals, and the antenna radiates the energy from the current as electromagnetic waves (radio waves). In reception, an antenna intercepts some of the power of an electromagnetic wave in order to produce a tiny voltage at its terminals, that is applied to a receiver to be amplified. Some antennas can be used for both transmitting and receiving, even simultaneously, depending on the connected equipment.

Propagation

Once generated, electromagnetic waves travel through space either directly, or have their path altered by reflection, refraction or diffraction. The intensity of the waves diminishes due to geometric dispersion (the inverse-square law); some energy may also be absorbed by the intervening medium in some cases. Noise will generally alter the desired signal; this electromagnetic interference comes from natural sources, as well as from artificial sources such as other transmitters and accidental radiators. Noise is also produced at every step due to the inherent properties of the devices used. If the magnitude of the noise is large enough, the desired signal will no longer be discernible; the signal-to-noise ratio is the fundamental limit to the range of radio communications.

Resonance

Electrical resonance of tuned circuits in radios allow individual stations to be selected. A resonant circuit will respond strongly to a particular frequency, and much less so to differing frequencies. This allows the radio receiver to discriminate between multiple signals differing in frequency.

Receiver and Demodulation

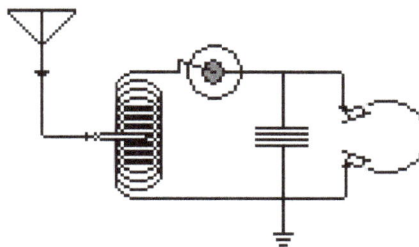

A crystal receiver, consisting of an antenna, adjustable electromagnetic coil, crystal rectifier, capacitor, headphones and ground connection.

The electromagnetic wave is intercepted by a tuned receiving antenna; this structure captures some of the energy of the wave and returns it to the form of oscillating electrical currents. At the receiver, these currents are demodulated, which is conversion to a usable signal form by a detector sub-system. The receiver is "tuned" to respond preferentially to the desired signals, and reject undesired signals.

Early radio systems relied entirely on the energy collected by an antenna to produce signals for the operator. Radio became more useful after the invention of electronic devices such as the vacuum tube and later the transistor, which made it possible to amplify weak signals. Today radio systems are used for applications from walkie-talkie children's toys to the control of space vehicles, as well as for broadcasting, and many other applications.

A *radio receiver* receives its input from an antenna, uses electronic filters to separate a wanted radio signal from all other signals picked up by this antenna, amplifies it to a level suitable for further processing, and finally converts through demodulation and decoding the signal into a form usable for the consumer, such as sound, pictures, digital data, measurement values, navigational positions, etc.

Radio Band

Light comparison		
Name	**Frequency (Hz) (Wavelength)**	**Photon energy (eV)**
Gamma ray	> 30 EHz (0.01 nm)	124 keV - 300+ GeV
X-Ray	30 EHz - 30 PHz (0.01 nm - 10 nm)	124 eV to 120 keV
Ultraviolet	30 PHz - 750 THz (10 nm - 400 nm)	3.1 eV to 124 eV
Visible	750 THz - 428.5 THz (400 nm - 700 nm)	1.7 eV - 3.1 eV
Infrared	428.5 THz - 300 GHz (700 nm - 1 mm)	1.24 meV - 1.7 eV
Microwave	300 GHz - 300 MHz (1 mm - 1 m)	1.24 μeV - 1.24 meV
Radio	300 MHz - 3 kHz (1 m - 100 km)	12.4 feV - 1.24 meV

Radio frequencies occupy the range from a 3 kHz to 300 GHz, although commercially important uses of radio use only a small part of this spectrum. Other types of electromagnetic radiation, with frequencies above the RF range, are infrared, visible light, ultraviolet, X-rays and gamma rays. Since the energy of an individual photon of radio frequency is too low to remove an electron from an atom, radio waves are classified as non-ionizing radiation.

Communication Systems

A *radio communication system* sends signals by radio. Types of radio communication systems deployed depend on technology, standards, regulations, radio spectrum allocation, user requirements, service positioning, and investment.

The radio equipment involved in communication systems includes a transmitter and a receiver, each having an antenna and appropriate terminal equipment such as a microphone at the transmitter and a loudspeaker at the receiver in the case of a voice-communication system. The power consumed in a transmitting station varies depending on the distance of communication and the transmission conditions. The power received at the receiving station is usually only a tiny fraction of the transmitter's output, since communication depends on receiving the information, not the energy, that was transmitted.

Classical radio communications systems use frequency-division multiplexing (FDM) as a strategy to split up and share the available radio-frequency bandwidth for use by different parties' communications concurrently. Modern radio communication systems include those that divide up a radio-frequency band by time-division multiplexing (TDM) and code-division multiplexing (CDM) as alternatives to the classical FDM strategy. These systems offer different tradeoffs in supporting multiple users, beyond the FDM strategy that was ideal for broadcast radio but less so for applications such as mobile telephony.

A radio communication system may send information only one way. For example, in broadcasting a single transmitter sends signals to many receivers. Two stations may take turns sending and receiving, using a single radio frequency; this is called "simplex." By using two radio frequencies, two stations may continuously and concurrently send and receive signals - this is called "duplex" operation.

History

In 1864 James Clerk Maxwell showed mathematically that electromagnetic waves could propagate through free space. The effects of electromagnetic waves (then-unexplained "action at a distance" sparking behavior) were actually observed before and after Maxwell's work by many inventors and experimenters including George Adams (1780-1784), Luigi Galvani (1791), Peter Samuel Munk (1835), Joseph Henry (1842), Samuel Alfred Varley (1852), Edwin Houston, Elihu Thomson, Thomas Edison (1875) and David Edward Hughes (1878). Edison gave the effect the name "etheric force" and Hughes detected a spark impulse up to 500 yards (460 m) with a portable receiver, but none could identify what caused the phenomenon and it was usually written off as electromagnetic induction. In 1886 Heinrich Rudolf Hertz noticed the same sparking phenomenon and, in published experiments (1887-1888), was able to demonstrate the existence of electromagnetic waves in an experiment confirming Maxwell's theory of electromagnetism.

The discovery of these "Hertzian waves" (radio waves) prompted many experiments by physicists. An August 1894 lecture by the British physicist Oliver Lodge, where he transmitted and received "Hertzian waves" at distances up to 50 meters, was followed up the same year with experiments by Indian physicist Jagadish Chandra Bose in extremely high frequency radio microwave optics and a year later with the construction of a radio based lightning detector by Russian physicist Alexander Stepanovich Popov. Starting in late 1894, Guglielmo Marconi began pursuing the idea of building a wireless telegraphy system based on Hertzian waves (radio). Marconi gained a patent on the system in 1896 and developed it into a commercial communication system over the next few years.

Early 20th century radio systems transmitted messages by continuous wave code only. Early attempts at developing a system of amplitude modulation for voice and music were demonstrated in 1900 and 1906, but had little success. World War I accelerated the development of radio for military communications, and in this era the first vacuum tubes were applied to radio transmitters and receivers. Electronic amplification was a key development in changing radio from an experimental practice by experts into a home appliance. After the war, commercial radio broadcasting began in the 1920s and became an important mass medium for entertainment and news. World War II again accelerated development of radio for the wartime purposes of aircraft and land communication, radio navigation and radar. After the war, the experiments in television that had been interrupted were resumed, and it also became an important home entertainment medium.

Uses of Radio

Early uses were maritime, for sending telegraphic messages using Morse code between ships and land. The earliest users included the Japanese Navy scouting the Russian fleet during the Battle of Tsushima in 1905. One of the most memorable uses of marine telegraphy was during the sinking of the RMS *Titanic* in 1912, including communications between operators on the sinking ship and nearby vessels, and communications to shore stations listing the survivors.

Radio was used to pass on orders and communications between armies and navies on both sides in World War I; Germany used radio communications for diplomatic messages once it discovered that its submarine cables had been tapped by the British. The United States passed on President Woodrow Wilson's Fourteen Points to Germany via radio during the war. Broadcasting began from San Jose, California in 1909, and became feasible in the 1920s, with the widespread introduction of radio receivers, particularly in Europe and the United States. Besides broadcasting, point-to-point broadcasting, including telephone messages and relays of radio programs, became widespread in the 1920s and 1930s. Another use of radio in the pre-war years was the development of detection and locating of aircraft and ships by the use of radar (*RA*dio *D*etection *A*nd *R*anging).

Today, radio takes many forms, including wireless networks and mobile communications of all types, as well as radio broadcasting. Before the advent of television, commercial radio broadcasts included not only news and music, but dramas, comedies, variety shows, and many other forms of entertainment (the era from the late 1920s to the mid-1950s is commonly called radio's "Golden Age"). Radio was unique among methods of dramatic presentation in that it used only sound.

Audio

One-way

AM radio uses amplitude modulation, in which the amplitude of the transmitted signal is made proportional to the sound amplitude captured (transduced) by the microphone, while the transmitted frequency remains unchanged. Transmissions are affected by static and interference because lightning and other sources of radio emissions on the same frequency add their amplitudes to the original transmitted amplitude.

Bakelite radio at the Bakelite Museum, Orchard Mill, Williton, Somerset, UK.

A Fisher 500 AM/FM hi-fi receiver from 1959.

In the early part of the 20th century, American AM radio stations broadcast with powers as high as 500 kW, and some could be heard worldwide; these stations' transmitters were commandeered for military use by the US Government during World War II. Currently, the maximum broadcast power for a civilian AM radio station in the United States and Canada is 50 kW. In 1986 KTNN received the last granted 50,000-watt class A license. These 50 kW stations are generally called "clear channel" stations, because within North America each of these stations has exclusive use of its broadcast frequency throughout part or all of the broadcast day.

Bush House, old home of the BBC World Service.

FM broadcast radio sends music and voice with less noise than AM radio. It is often mistakenly thought that FM is higher fidelity than AM, but that is not true. AM is capable of the same audio bandwidth that FM employs. AM receivers typically use narrower filters in the receiver to recover the signal with less noise. AM stereo receivers can reproduce the same audio bandwidth that FM does due to the wider filter used in an AM stereo receiver, but today, AM radios limit the audio bandpass to 3–5 kHz. In frequency modulation, amplitude variation at the microphone causes the transmitter frequency to fluctuate. Because the audio signal modulates the frequency and not the amplitude, an FM signal is not subject to static and interference in the same way as AM signals. Due to its need for a wider bandwidth, FM is transmitted in the Very High Frequency (VHF, 30 MHz to 300 MHz) radio spectrum.

VHF radio waves act more like light, traveling in straight lines; hence the reception range is generally limited to about 50–200 miles (80–322 km). During unusual upper atmospheric conditions, FM signals are occasionally reflected back towards the Earth by the ionosphere, resulting in long distance FM reception. FM receivers are subject to the capture effect, which causes the radio to only receive the strongest signal when multiple signals appear on the same frequency. FM receivers are relatively immune to lightning and spark interference.

High power is useful in penetrating buildings, diffracting around hills, and refracting in the dense atmosphere near the horizon for some distance beyond the horizon. Consequently, 100,000-watt FM stations can regularly be heard up to 100 miles (160 km) away, and farther, 150 miles (240 km), if there are no competing signals. A few old, "grandfathered" stations do not conform to these power rules. WBCT-FM (93.7) in Grand Rapids, Michigan, US, runs 320,000 watts ERP, and can increase to 500,000 watts ERP by the terms of its original license. Such a huge power level does not usually help to increase range as much as one might expect, because VHF frequencies travel in nearly straight lines over the horizon and off into space.

FM subcarrier services are secondary signals transmitted in a "piggyback" fashion along with the main program. Special receivers are required to utilize these services. Analog channels may contain alternative programming, such as reading services for the blind, background music or stereo sound signals. In some extremely crowded metropolitan areas, the sub-channel program might be an alternate foreign-language radio program for various ethnic groups. Sub-carriers can also transmit digital data, such as station identification, the current song's name, web addresses, or stock quotes. In some countries, FM radios automatically re-tune themselves to the same channel in a different district by using sub-bands.

Two-way

Aviation voice radios use Aircraft band VHF AM. AM is used so that multiple stations on the same channel can be received. (Use of FM would result in stronger stations blocking out reception of weaker stations due to FM's capture effect). Aircraft fly high enough that their transmitters can be received hundreds of miles away, even though they are using VHF.

Degen DE1103, an advanced world mini-receiver with single sideband modulation and dual conversion

Marine voice radios can use single sideband voice (SSB) in the shortwave High Frequency (HF—3 MHz to 30 MHz) radio spectrum for very long ranges or Marine VHF radio / *narrowband FM* in the VHF spectrum for much shorter ranges. Narrowband FM sacrifices fidelity to make more channels available within the radio spectrum, by using a smaller range of radio frequencies, usually with five kHz of deviation, versus the 75 kHz used by commercial FM broadcasts, and 25 kHz used for TV sound.

Government, police, fire and commercial voice services also use narrowband FM on special frequencies. Early police radios used AM receivers to receive one-way dispatches. Civil and military HF (high frequency) voice services use shortwave radio to contact ships at sea, aircraft and isolated settlements. Most use single sideband voice (SSB), which uses less bandwidth than AM. On an AM radio SSB sounds like ducks quacking, or the adults in a Charlie Brown cartoon. Viewed as a graph of frequency versus power, an AM signal shows power where the frequencies of the voice add and subtract with the main radio frequency. SSB cuts the bandwidth in half by suppressing the carrier and one of the sidebands. This also makes the transmitter about three times more powerful, because it doesn't need to transmit the unused carrier and sideband.

TETRA, Terrestrial Trunked Radio is a digital cell phone system for military, police and ambulances. Commercial services such as XM, WorldSpace and Sirius offer encrypted digital satellite radio.

Telephony

Mobile phones transmit to a local cell site (transmitter/receiver) that ultimately connects to the public switched telephone network (PSTN) through an optic fiber or microwave radio and other network elements. When the mobile phone nears the edge of the cell site's radio coverage area, the central computer switches the phone to a new cell. Cell phones originally used FM, but now most use either GSM or CDMA digital modulation schemes. Satellite phones use satellites rather than cell towers to communicate.

Video

Analog television sends the picture as AM and the sound as AM or FM, with the sound carrier a fixed frequency (4.5 MHz in the NTSC system) away from the video carrier. Analog television also uses a vestigial sideband on the video carrier to reduce the bandwidth required.

Digital television uses 8VSB modulation in North America (under the ATSC digital television standard), and COFDM modulation elsewhere in the world (using the DVB-T standard). A Reed–Solomon error correction code adds redundant correction codes and allows reliable reception during moderate data loss. Although many current and future codecs can be sent in the MPEG transport stream container format, as of 2006 most systems use a standard-definition format almost identical to DVD: MPEG-2 video in Anamorphic widescreen and MPEG layer 2 (*MP2*) audio. High-definition television is possible simply by using a higher-resolution picture, but H.264/AVC is being considered as a replacement video codec in some regions for its improved compression. With the compression and improved modulation involved, a single "channel" can contain a high-definition program and several standard-definition programs.

Navigation

All satellite navigation systems use satellites with precision clocks. The satellite transmits its position, and the time of the transmission. The receiver listens to four satellites, and can figure its position as being on a line that is tangent to a spherical shell around each satellite, determined by the time-of-flight of the radio signals from the satellite. A computer in the receiver does the math.

Radio direction-finding is the oldest form of radio navigation. Before 1960 navigators used movable loop antennas to locate commercial AM stations near cities. In some cases they used marine radiolocation beacons, which share a range of frequencies just above AM radio with amateur radio operators. LORAN systems also used time-of-flight radio signals, but from radio stations on the ground.

Very High Frequency Omnidirectional Range (VOR), systems (used by aircraft), have an antenna array that transmits two signals simultaneously. A directional signal rotates like a lighthouse at a fixed rate. When the directional signal is facing north, an omnidirectional signal pulses. By measuring the difference in phase of these two signals, an aircraft can determine its bearing or radial from the station, thus establishing a line of position. An aircraft can get readings from two VORs and locate its position at the intersection of the two radials, known as a "fix."

When the VOR station is collocated with DME (Distance Measuring Equipment), the aircraft can determine its bearing and range from the station, thus providing a fix from only one ground station. Such stations are called VOR/DMEs. The military operates a similar system of navaids, called TACANs, which are often built into VOR stations. Such stations are called VORTACs. Because TACANs include distance measuring equipment, VOR/DME and VORTAC stations are identical in navigation potential to civil aircraft.

Radar

Radar (Radio Detection And Ranging) detects objects at a distance by bouncing radio waves off them. The delay caused by the echo measures the distance. The direction of the beam determines the direction of the reflection. The polarization and frequency of the return can sense the type of surface. Navigational radars scan a wide area two to four times per minute. They use very short waves that reflect from earth and stone. They are common on commercial ships and long-distance commercial aircraft.

General purpose radars generally use navigational radar frequencies, but modulate and polarize the pulse so the receiver can determine the type of surface of the reflector. The best general-purpose radars distinguish the rain of heavy storms, as well as land and vehicles. Some can superimpose sonar data and map data from GPS position.

Search radars scan a wide area with pulses of short radio waves. They usually scan the area two to four times a minute. Sometimes search radars use the Doppler effect to separate moving vehicles from clutter. Targeting radars use the same principle as search radar but scan a much smaller area far more often, usually several times a second or more. Weather radars resemble search radars, but use radio waves with circular polarization and a wavelength to reflect from water droplets. Some weather radar use the Doppler effect to measure wind speeds.

Data (Digital Radio)

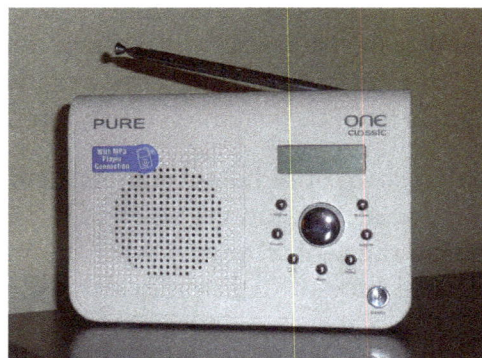

2008 Pure One Classic digital radio

Most new radio systems are digital, including Digital TV, satellite radio, and Digital Audio Broadcasting. The oldest form of digital broadcast was spark gap telegraphy, used by pioneers such as Marconi. By pressing the key, the operator could send messages in Morse code by energizing a rotating commutating spark gap. The rotating commutator produced a tone in the receiver, where a simple spark gap would produce a hiss, indistinguishable from static. Spark-gap transmitters are now illegal, because their transmissions span several hundred megahertz. This is very wasteful of both radio frequencies and power.

The next advance was continuous wave telegraphy, or CW (Continuous Wave), in which a pure radio frequency, produced by a vacuum tube electronic oscillator was switched on and off by a key. A receiver with a local oscillator would "heterodyne" with the pure radio frequency, creating a whistle-like audio tone. CW uses less than 100 Hz of bandwidth. CW is still used, these days primarily by amateur radio operators (hams). Strictly, on-off keying of a carrier should be known as "Interrupted Continuous Wave" or ICW or on-off keying (OOK).

Radioteletype equipment usually operates on short-wave (HF) and is much loved by the military because they create written information without a skilled operator. They send a bit as one of two tones using frequency-shift keying. Groups of five or seven bits become a character printed by a teleprinter. From about 1925 to 1975, radioteletype was how most commercial messages were sent to less developed countries. These are still used by the military and weather services.

Aircraft use a 1200 Baud radioteletype service over VHF to send their ID, altitude and position, and get gate and connecting-flight data. Microwave dishes on satellites, telephone exchanges and TV stations usually use quadrature amplitude modulation (QAM). QAM sends data by changing both the phase and the amplitude of the radio signal. Engineers like QAM because it packs the most bits into a radio signal when given an exclusive (non-shared) fixed narrowband frequency range. Usually the bits are sent in "frames" that repeat. A special bit pattern is used to locate the beginning of a frame.

Modern GPS receivers.

Communication systems that limit themselves to a fixed narrowband frequency range are vulnerable to jamming. A variety of jamming-resistant spread spectrum techniques were initially developed for military use, most famously for Global Positioning System satellite transmissions. Commercial use of spread spectrum began in the 1980s. Bluetooth, most cell phones, and the 802.11b version of Wi-Fi each use various forms of spread spectrum.

Systems that need reliability, or that share their frequency with other services, may use "coded orthogonal frequency-division multiplexing" or COFDM. COFDM breaks a digital signal into as many as several hundred slower subchannels. The digital signal is often sent as QAM on the subchannels. Modern COFDM systems use a small computer to make and decode the signal with digital signal processing, which is more flexible and far less expensive than older systems that implemented separate electronic channels.

COFDM resists fading and ghosting because the narrow-channel QAM signals can be sent slowly. An adaptive system, or one that sends error-correction codes can also resist interference, because most interference can affect only a few of the QAM channels. COFDM is used for Wi-Fi, some cell phones, Digital Radio Mondiale, Eureka 147, and many other local area network, digital TV and radio standards.

Heating

Radio-frequency energy generated for heating of objects is generally not intended to radiate outside of the generating equipment, to prevent interference with other radio signals. Microwave ovens use intense radio waves to heat food. Diathermy equipment is used in surgery for sealing of blood vessels. Induction furnaces are used for melting metal for casting, and induction hobs for cooking.

Amateur Radio Service

Amateur radio station with multiple receivers and transceivers

Amateur radio, also known as "ham radio", is a hobby in which enthusiasts are licensed to communicate on a number of bands in the radio frequency spectrum non-commercially and for their own experiments. They may also provide emergency and service assistance in exceptional circumstances. This contribution has been very beneficial in saving lives in many instances.

Radio amateurs use a variety of modes, including efficient ones like Morse code and experimental ones like Low-Frequency Experimental Radio. Several forms of radio were pioneered by radio amateurs and later became commercially important, including FM, single-sideband (SSB), AM, digital packet radio and satellite repeaters. Some amateur frequencies may be disrupted illegally by power-line internet service.

Unlicensed Radio Services

Unlicensed, government-authorized personal radio services such as Citizens' band radio in Australia, most of the Americas, and Europe, and Family Radio Service and Multi-Use Radio Service in North America exist to provide simple, usually short range communication for individuals and small groups, without the overhead of licensing. Similar services exist in other parts of the world. These radio services involve the use of handheld units.

Wi-Fi also operates in unlicensed radio bands and is very widely used to network computers.

Free radio stations, sometimes called pirate radio or "clandestine" stations, are unauthorized, unlicensed, illegal broadcasting stations. These are often low power transmitters operated on sporadic schedules by hobbyists, community activists, or political and cultural dissidents. Some pirate stations operating offshore in parts of Europe and the United Kingdom more closely resembled legal stations, maintaining regular schedules, using high power, and selling commercial advertising time.

Radio Control (RC)

Radio remote controls use radio waves to transmit control data to a remote object as in some early forms of guided missile, some early TV remotes and a range of model boats, cars and airplanes. Large industrial remote-controlled equipment such as cranes and switching locomotives now usually use digital radio techniques to ensure safety and reliability.

In Madison Square Garden, at the Electrical Exhibition of 1898, Nikola Tesla successfully demonstrated a radio-controlled boat. He was awarded U.S. patent No. 613,809 for a "Method of and Apparatus for Controlling Mechanism of Moving Vessels or Vehicles."

Software-defined Radio

Software-defined radio (SDR) is a radio communication system where components that have been typically implemented in hardware (e.g. mixers, filters, amplifiers, modulators/demodulators, detectors, etc.) are instead implemented by means of software on a personal computer or embedded system. While the concept of SDR is not new, the rapidly evolving capabilities of digital electronics render practical many processes which used to be only theoretically possible.

Overview

A basic SDR system may consist of a personal computer equipped with a sound card, or other analog-to-digital converter, preceded by some form of RF front end. Significant amounts of signal processing are handed over to the general-purpose processor, rather than being done in special-purpose hardware (electronic circuits). Such a design produces a radio which can receive and transmit widely different radio protocols (sometimes referred to as waveforms) based solely on the software used.

Software radios have significant utility for the military and cell phone services, both of which must serve a wide variety of changing radio protocols in real time.

In the long term, software-defined radios are expected by proponents like the SDRForum (now The Wireless Innovation Forum) to become the dominant technology in radio communications. SDRs, along with software defined antennas are the enablers of the cognitive radio.

A software-defined radio can be flexible enough to avoid the "limited spectrum" assumptions of designers of previous kinds of radios, in one or more ways including:

- Spread spectrum and ultrawideband techniques allow several transmitters to transmit in the same place on the same frequency with very little interference, typically combined with one or more error detection and correction techniques to fix all the errors caused by that interference.

- Software defined antennas adaptively "lock onto" a directional signal, so that receivers can better reject interference from other directions, allowing it to detect fainter transmissions.

- Cognitive radio techniques: each radio measures the spectrum in use and communicates that information to other cooperating radios, so that transmitters can avoid mutual interference by selecting unused frequencies. Alternatively, each radio connects to a geolocation database to obtain information about the spectrum occupancy in its location and, flexibly, adjusts its operating frequency and/or transmit power not to cause interference to other wireless services.

- Dynamic transmitter power adjustment, based on information communicated from the receivers, lowering transmit power to the minimum necessary, reducing the near-far problem and reducing interference to others, and extending battery life in portable equipment.

- Wireless mesh network where every added radio increases total capacity and reduces the power required at any one node. Each node only transmits loudly enough for the message to hop to the nearest node in that direction, reducing near-far problem and reducing interference to others.

Operating Principles

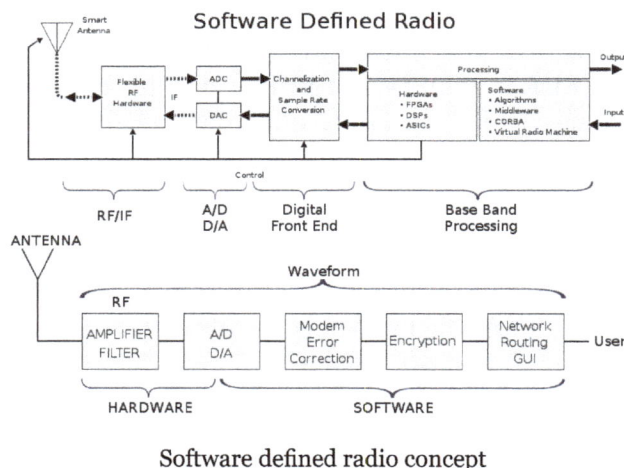

Software defined radio concept

Ideal Concept

The ideal receiver scheme would be to attach an analog-to-digital converter to an antenna. A digital signal processor would read the converter, and then its software would transform the stream of data from the converter to any other form the application requires.

An ideal transmitter would be similar. A digital signal processor would generate a stream of numbers. These would be sent to a digital-to-analog converter connected to a radio antenna.

The ideal scheme is not completely realizable due to the actual limits of the technology. The main problem in both directions is the difficulty of conversion between the digital and the analog domains at a high enough rate and a high enough accuracy at the same time, and without relying upon physical processes like interference and electromagnetic resonance for assistance.

Receiver Architecture

Most receivers use a variable-frequency oscillator, mixer, and filter to tune the desired signal to a common intermediate frequency or baseband, where it is then sampled by the analog-to-digital converter. However, in some applications it is not necessary to tune the signal to an intermediate frequency and the radio frequency signal is directly sampled by the analog-to-digital converter (after amplification).

Real analog-to-digital converters lack the dynamic range to pick up sub-microvolt, nanowatt-power radio signals. Therefore, a low-noise amplifier must precede the conversion step and this device introduces its own problems. For example, if spurious signals are present (which is typical), these compete with the desired signals within the amplifier's dynamic range. They may introduce distortion in the desired signals, or may block them completely. The standard solution is to put band-pass filters between the antenna and the amplifier, but these reduce the radio's flexibility. Real software radios often have two or three analog channel filters with different bandwidths that are switched in and out.

History

The term "digital receiver" was coined in 1970 by a researcher at a United States Department of Defense laboratory. A laboratory called the Gold Room at TRW in California created a software baseband analysis tool called Midas, which had its operation defined in software.

The term "software radio" was coined in 1984 by a team at the Garland, Texas Division of E-Systems Inc. (now Raytheon) to refer to a digital baseband receiver and published in their E-Team company newsletter. A 'Software Radio Proof-of-Concept' laboratory was developed there that popularized Software Radio within various government agencies. This 1984 Software Radio was a digital baseband receiver that provided programmable interference cancellation and demodulation for broadband signals, typically with thousands of adaptive filter taps, using multiple array processors accessing shared memory.

In 1991, Joe Mitola independently reinvented the term software radio for a plan to build a GSM base station that would combine Ferdensi's digital receiver with E-Systems Melpar's digitally controlled communications jammers for a true software-based transceiver. E-Systems Melpar sold the software radio idea to the US Air Force. Melpar built a prototype commanders' tactical terminal in 1990-91 that employed Texas Instruments TMS320C30 processors and Harris digital receiver chip sets with digitally synthesized transmission. That prototype didn't last long because when E-Systems ECI Division manufactured the first limited production units, they decided to "throw out those useless C30 boards," replacing them with conventional RF filtering on transmit and receive, reverting to a digital baseband radio instead of the SPEAKeasy like IF ADC/DACs of Mitola's prototype. The Air Force would not let Mitola publish the technical details of that prototype, nor would they let Diane Wasserman publish related software life cycle lessons learned because they regarded it as a "USAF competitive advantage." So instead, with USAF permission, in 1991 Mitola described the architecture principles without implementation details in a paper, "Software Radio: Survey, Critical Analysis and Future Directions" which became the first IEEE publication to employ the term in 1992. When Mitola presented the paper at the conference, Bob Prill of GEC Marconi began his presentation following Mitola with "Joe is absolutely right about the theory of a software radio and we are building one." Prill gave a GEC Marconi paper on PAVE PILLAR, a SPEAKeasy precursor. SPEAKeasy, the military software radio was formulated by Wayne Bonser, then of Rome Air Development Center (RADC), now Rome Labs; by Alan Margulies of MITRE Rome, NY; and then Lt Beth Kaspar, the original DARPA SPEAKeasy project manager and by others at Rome including Don Upmal. Although Mitola's IEEE publications resulted in the largest global footprint for software radio, Mitola privately credits that DoD lab of the 1970s with its leaders Carl, Dave, and John with inventing the digital receiver technology on which he based software radio once it was possible to transmit via software.

A few months after the National Telesystems Conference 1992, in an E-Systems corporate program review, a vice-president of E-Systems Garland Division objected to Melpar's (Mitola's) use of the term "software radio" without credit to Garland. Alan Jackson, Melpar VP of marketing at that time asked the Garland VP if their laboratory or devices included transmitters. The Garland VP said "No, of course not — ours is a software radio receiver". Al replied "Then it's a digital receiver but without a transmitter, it's not a software radio." Corporate leadership agreed with Al, so the publication stood. Many amateur radio operators and HF radio engineers had realized the value of digitizing HF at RF and of processing it with Texas Instruments TI C30 digital signal processors (DSPs) and their precursors during the 1980s and early 1990s. Radio engineers at Roke Manor in the UK and at an organization in Germany had recognized the benefits of ADC at the RF in parallel, so success has many fathers. Mitola's publication of software radio in the IEEE opened the concept to the broad community of radio engineers. His landmark May 1995 special issue of the IEEE Communications Magazine with the cover "Software Radio" was widely regarded as watershed event with thousands of academic citations. Mitola was introduced by Joao da Silva in 1997 at the First International Conference on Software Radio as "godfather" of software radio in no small part for his willingness to share such a valuable technology "in the public interest."

Perhaps the first software-based radio transceiver was designed and implemented by Peter Hoeher and Helmuth Lang at the German Aerospace Research Establishment (DLR, formerly DFVLR) in Oberpfaffenhofen, Germany, in 1988. Both transmitter and receiver of an adaptive digital satellite modem were implemented according to the principles of a software radio, and a flexible hardware periphery was proposed.

The term "software defined radio" was coined in 1995 by Stephen Blust, who published a request for information from Bell South Wireless at the first meeting of the Modular Multifunction Information Transfer Systems (MMITS) forum in 1996, organized by the USAF and DARPA around the commercialization of their SPEAKeasy II program. Mitola objected to Blust's term, but finally accepted it as a pragmatic pathway towards the ideal software radio. Though the concept was first implemented with an IF ADC in the early 1990s, software-defined radios have their origins in the defense sector since the late 1970s in both the U.S. and Europe (for example, Walter Tuttlebee described a VLF radio that used an ADC and an 8085 microprocessor). about a year after the First International Conference in Brussels. One of the first public software radio initiatives was the U.S. DARPA-Air Force military project named SpeakEasy. The primary goal of the SpeakEasy project was to use programmable processing to emulate more than 10 existing military radios, operating in frequency bands between 2 and 2000 MHz. Another SPEAKeasy design goal was to be able to easily incorporate new coding and modulation standards in the future, so that military communications can keep pace with advances in coding and modulation techniques.

SPEAKeasy Phase I

From 1990 to 1995, the goal of the SPEAKeasy program was to demonstrate a radio for the U.S. Air Force tactical ground air control party that could operate from 2 MHz to 2 GHz, and thus could interoperate with ground force radios (frequency-agile VHF, FM, and SINCGARS), Air Force radios (VHF AM), Naval Radios (VHF AM and HF SSB teleprinters) and satellites (microwave QAM). Some particular goals were to provide a new signal format in two weeks from a standing start, and demonstrate a radio into which multiple contractors could plug parts and software.

The project was demonstrated at TF-XXI Advanced Warfighting Exercise, and demonstrated all of these goals in a non-production radio. There was some discontent with failure of these early software radios to adequately filter out of band emissions, to employ more than the simplest of interoperable modes of the existing radios, and to lose connectivity or crash unexpectedly. Its cryptographic processor could not change context fast enough to keep several radio conversations on the air at once. Its software architecture, though practical enough, bore no resemblance to any other. The SPEAKeasy architecture was refined at the MMITS Forum between 1996 and 1999 and inspired the DoD integrated process team (IPT) for programmable modular communications systems (PMCS) to proceed with what became the Joint Tactical Radio System (JTRS).

The basic arrangement of the radio receiver used an antenna feeding an amplifier and down-converter feeding an automatic gain control, which fed an analog to digital converter that was on a computer VMEbus with a lot of digital signal processors (Texas Instruments C40s). The transmitter had digital to analog converters on the PCI bus feeding an up converter (mixer) that led to a power amplifier and antenna. The very wide frequency range was divided into a few sub-bands with different analog radio technologies feeding the same analog to digital converters. This has since become a standard design scheme for wide band software radios.

SPEAKeasy Phase II

The goal was to get a more quickly reconfigurable architecture, *i.e.*, several conversations at once, in an *open* software architecture, with cross-channel connectivity (the radio can "bridge" different radio protocols). The secondary goals were to make it smaller, cheaper, and weigh less.

The project produced a demonstration radio only fifteen months into a three-year research project. The demonstration was so successful that further development was halted, and the radio went into production with only a 4 MHz to 400 MHz range.

The software architecture identified standard interfaces for different modules of the radio: "radio frequency control" to manage the analog parts of the radio, "modem control" managed resources for modulation and demodulation schemes (FM, AM, SSB, QAM, etc.), "waveform processing" modules actually performed the modem functions, "key processing" and "cryptographic processing" managed the cryptographic functions, a "multimedia" module did voice processing, a "human interface" provided local or remote controls, there was a "routing" module for network services, and a "control" module to keep it all straight.

The modules are said to communicate without a central operating system. Instead, they send messages over the PCI computer bus to each other with a layered protocol.

As a military project, the radio strongly distinguished "red" (unsecured secret data) and "black" (cryptographically-secured data).

The project was the first known to use FPGAs (field programmable gate arrays) for digital processing of radio data. The time to reprogram these was an issue limiting application of the radio. Today, the time to write a program for an FPGA is still significant, but the time to download a stored FPGA program is around 20 milliseconds. This means an SDR could change transmission protocols and frequencies in one fiftieth of a second, probably not an intolerable interruption for that task.

Military

USA

The Joint Tactical Radio System (JTRS) was a program of the US military to produce radios that provide flexible and interoperable communications. Examples of radio terminals that require support include hand-held, vehicular, airborne and dismounted radios, as well as base-stations (fixed and maritime).

This goal is achieved through the use of SDR systems based on an internationally endorsed open Software Communications Architecture (SCA). This standard uses CORBA on POSIX operating systems to coordinate various software modules.

The program is providing a flexible new approach to meet diverse soldier communications needs through software programmable radio technology. All functionality and expandability is built upon the SCA.

The SCA, despite its military origin, is under evaluation by commercial radio vendors for applicability in their domains. The adoption of general-purpose SDR frameworks outside of military, intelligence, experimental and amateur uses, however, is inherently hampered by the fact that civilian users can more easily settle with a fixed architecture, optimized for a specific function, and as such more economical in mass market applications. Still, software defined radio's inherent flexibility can yield substantial benefits in the longer run, once the fixed costs of implementing it have gone down enough to overtake the cost of iterated redesign of purpose built systems. This then explains the increasing commercial interest in the technology.

SCA-based infrastructure software and rapid development tools for SDR education and research are provided by the Open Source SCA Implementation – Embedded (OSSIE) project. The Wireless

Innovation Forum funded the SCA Reference Implementation project, an open source implementation of the SCA specification. (SCARI) can be downloaded for free.

Amateur and Home use

Microtelecom Perseus - a HF SDR for the amateur radio market

A typical amateur software radio uses a direct conversion receiver. Unlike direct conversion receivers of the more distant past, the mixer technologies used are based on the quadrature sampling detector and the quadrature sampling exciter.

The receiver performance of this line of SDRs is directly related to the dynamic range of the analog-to-digital converters (ADCs) utilized. Radio frequency signals are down converted to the audio frequency band, which is sampled by a high performance audio frequency ADC. First generation SDRs used a PC sound card to provide ADC functionality. The newer software defined radios use embedded high performance ADCs that provide higher dynamic range and are more resistant to noise and RF interference.

A fast PC performs the digital signal processing (DSP) operations using software specific for the radio hardware. Several software radio efforts use the open source SDR library DttSP.

The SDR software performs all of the demodulation, filtering (both radio frequency and audio frequency), and signal enhancement (equalization and binaural presentation). Uses include every common amateur modulation: morse code, single sideband modulation, frequency modulation, amplitude modulation, and a variety of digital modes such as radioteletype, slow-scan television, and packet radio. Amateurs also experiment with new modulation methods: for instance, the DREAM open-source project decodes the COFDM technique used by Digital Radio Mondiale.

There is a broad range of hardware solutions for radio amateurs and home use. There are professional-grade transceiver solutions, e.g. the Zeus ZS-1 or the Flex Radio, home-brew solutions,e.g. PicAStar transceiver, the SoftRock SDR kit, and starter or professional receiver solutions, e.g. the FiFi SDR for shortwave, or the Quadrus coherent multi-channel SDR receiver for short wave or VHF/UHF in direct digital mode of operation.

Internals of a low-cost DVB-T USB dongle that uses Realtek RTL2832U (square IC on the right) as the controller and Rafael Micro R820T (square IC on the left) as the tuner.

It has been discovered that some common low-cost DVB-T USB dongles with the Realtek RTL2832U controller and tuner, e.g. the Elonics E4000 or the Rafael Micro R820T, can be used as a wide-band SDR receiver. Recent experiments have proven the capability of this setup to analyze perseids shower using the graves radar signals.

GNU Radio logo

More recently, the GNU Radio using primarily the Universal Software Radio Peripheral (USRP) uses a USB 2.0 interface, an FPGA, and a high-speed set of analog-to-digital and digital-to-analog converters, combined with reconfigurable free software. Its sampling and synthesis bandwidth is a thousand times that of PC sound cards, which enables wideband operation.

The HPSDR (High Performance Software Defined Radio) project uses a 16-bit 135 MSPS analog-to-digital converter that provides performance over the range 0 to 55 MHz comparable to that of a conventional analogue HF radio. The receiver will also operate in the VHF and UHF range using either mixer image or alias responses. Interface to a PC is provided by a USB 2.0 interface, although Ethernet could be used as well. The project is modular and comprises a backplane onto which other boards plug in. This allows experimentation with new techniques and devices without the need to replace the entire set of boards. An exciter provides 1/2 W of RF over the same range or into the VHF and UHF range using image or alias outputs.

WebSDR is a project initiated by Pieter-Tjerk de Boer providing access via browser to multiple SDR receivers worldwide covering the complete shortwave spectrum. Recently he has analyzed Chirp Transmitter signals using the coupled system of receivers.

Cognitive Radio

A cognitive radio (CR) is a radio that can be programmed and configured dynamically to use the best wireless channels in its vicinity. Such a radio automatically detects available channels in wireless spectrum, then accordingly changes its transmission or reception parameters to allow more concurrent wireless communications in a given spectrum band at one location. This process is a form of dynamic spectrum management.

Description

In response to the operator's commands, the cognitive engine is capable of configuring radio-system parameters. These parameters include "waveform, protocol, operating frequency, and networking". This functions as an autonomous unit in the communications environment, exchanging information about the environment with the networks it accesses and other cognitive radios (CRs). A CR "monitors its own performance continuously", in addition to "reading the radio's outputs"; it then uses this information to "determine the RF environment, channel conditions, link performance, etc.", and adjusts the "radio's settings to deliver the required quality of service subject to an appropriate combination of user requirements, operational limitations, and regulatory constraints".

Some "smart radio" proposals combine wireless mesh network—dynamically changing the path messages take between two given nodes using cooperative diversity; cognitive radio—dynamically changing the frequency band used by messages between two consecutive nodes on the path; and software-defined radio—dynamically changing the protocol used by message between two consecutive nodes.

J. H. Snider, Lawrence Lessig, David Weinberger, and others say that low power "smart" radio is inherently superior to standard broadcast radio.

History

The concept of cognitive radio was first proposed by Joseph Mitola III in a seminar at KTH (the Royal Institute of Technology in Stockholm) in 1998 and published in an article by Mitola and Gerald Q. Maguire, Jr. in 1999. It was a novel approach in wireless communications, which Mitola later described as:

The point in which wireless personal digital assistants (PDAs) and the related networks are sufficiently computationally intelligent about radio resources and related computer-to-computer communications to detect user communications needs as a function of use context, and to provide radio resources and wireless services most appropriate to those needs.

Cognitive radio is considered as a goal towards which a software-defined radio platform should evolve: a fully reconfigurable wireless transceiver which automatically adapts its communication parameters to network and user demands.

Traditional regulatory structures have been built for an analog model and are not optimized for cognitive radio. Regulatory bodies in the world (including the Federal Communications Commission in the United States and Ofcom in the United Kingdom) as well as different independent measurement campaigns found that most radio frequency spectrum was inefficiently utilized. Cellular network bands are overloaded in most parts of the world, but other frequency bands (such as military, amateur radio and paging frequencies) are insufficiently utilized. Independent studies performed in some countries confirmed that observation, and concluded that spectrum utilization depends on time and place. Moreover, fixed spectrum allocation prevents rarely used frequencies (those assigned to specific services) from being used, even when any unlicensed users would not cause noticeable interference to the assigned service. Regulatory bodies in the world have been considering whether to allow unlicensed users in licensed bands if they would not cause any interference to licensed users. These initiatives have focused cognitive-radio research on dynamic spectrum access.

The first cognitive radio wireless regional area network standard, IEEE 802.22, was developed by IEEE 802 LAN/MAN Standard Committee (LMSC) and published in 2011. This standard uses geolocation and spectrum sensing for spectral awareness. Geolocation combines with a database of licensed transmitters in the area to identify available channels for use by the cognitive radio network. Spectrum sensing observes the spectrum and identifies occupied channels. IEEE 802.22 was designed to utilize the unused frequencies or fragments of time in a location. This white space is unused television channels in the geolocated areas. However, cognitive radio cannot occupy the same unused space all the time. As spectrum availability changes, the network adapts to prevent interference with licensed transmissions.

Terminology

Depending on transmission and reception parameters, there are two main types of cognitive radio:

- *Full Cognitive Radio* (Mitola radio), in which every possible parameter observable by a wireless node (or network) is considered.

- *Spectrum-Sensing Cognitive Radio*, in which only the radio-frequency spectrum is considered.

Other types are dependent on parts of the spectrum available for cognitive radio:

- *Licensed-Band Cognitive Radio*, capable of using bands assigned to licensed users (except for unlicensed bands, such as the U-NII band or the ISM band). The IEEE 802.22 working group is developing a standard for wireless regional area network (WRAN), which will operate on unused television channels, also known as TV white spaces.

- *Unlicensed-Band Cognitive Radio*, which can only utilize unlicensed parts of the radio frequency (RF) spectrum. One such system is described in the IEEE 802.15 Task Group 2 specifications, which focus on the coexistence of IEEE 802.11 and Bluetooth.

- *Spectrum mobility*: Process by which a cognitive-radio user changes its frequency of operation. Cognitive-radio networks aim to use the spectrum in a dynamic manner by allowing radio terminals to operate in the best available frequency band, maintaining seamless communication requirements during transitions to better spectrum.

- *Spectrum sharing*: Spectrum sharing cognitive radio networks allow cognitive radio users to share the spectrum bands of the licensed-band users. However, the cognitive radio users have to restrict their transmit power so that the interference caused to the licensed-band users is kept below a certain threshold.

- *Sensing-based Spectrum sharing*: In sensing-based spectrum sharing cognitive radio networks, cognitive radio users first listen to the spectrum allocated to the licensed users to detect the state of the licensed users. Based on the detection results, cognitive radio users decide their transmission strategies. If the licensed users are not using the bands, cognitive radio users will transmit over those bands. If the licensed users are using the bands, cognitive radio users share the spectrum bands with the licensed users by restricting their transmit power.

- *Database-enabled Spectrum Sharing,,*: In this modality of spectrum sharing, cognitive radio users are required to access a white space database prior to be allowed, or denied, access to the shared spectrum. The white space database contain algorithms, mathematical models and local regulations to predict the spectrum utilization in a geographical area and to infer on the risk of interference posed to incumbent services by a cognitive radio user accessing the shared spectrum. If the white space database judges that destructive interference to incumbents will happen, the cognitive radio user is denied access to the shared spectrum.

Technology

Although cognitive radio was initially thought of as a software-defined radio extension (full cognitive radio), most research work focuses on spectrum-sensing cognitive radio (particularly in the TV bands). The chief problem in spectrum-sensing cognitive radio is designing high-quality spectrum-sensing devices and algorithms for exchanging spectrum-sensing data between nodes. It has been shown that a simple energy detector cannot guarantee the accurate detection of signal presence, calling for more sophisticated spectrum sensing techniques and requiring information about spectrum sensing to be regularly exchanged between nodes. Increasing the number of cooperating sensing nodes decreases the probability of false detection.

Filling free RF bands adaptively, using OFDMA, is a possible approach. Timo A. Weiss and Friedrich K. Jondral of the University of Karlsruhe proposed a spectrum pooling system, in which free bands (sensed by nodes) were immediately filled by OFDMA subbands. Applications of spectrum-sensing cognitive radio include emergency-network and WLAN higher throughput and transmission-distance extensions. The evolution of cognitive radio toward cognitive networks is underway; the concept of cognitive networks is to intelligently organize a network of cognitive radios.

Functions

The main functions of cognitive radios are:

- *Power Control*: Power control is usually used for spectrum sharing CR systems to maximize the capacity of secondary users with interference power constraints to protect the primary users.

- *Spectrum sensing*: Detecting unused spectrum and sharing it, without harmful interference to other users; an important requirement of the cognitive-radio network is to sense empty spectrum. Detecting primary users is the most efficient way to detect empty spectrum. Spectrum-sensing techniques may be grouped into three categories:

o *Transmitter detection*: Cognitive radios must have the capability to determine if a signal from a primary transmitter is locally present in a certain spectrum. There are several proposed approaches to transmitter detection:

- Matched filter detection.

- Energy detection: Energy detection is a spectrum sensing method that detects the presence/absence of a signal just by measuring the received signal power. This signal detection approach is quite easy and convenient for practical implementation. To implement energy detector, however, noise variance information is required. It has been shown that an imperfect knowledge of the noise power (noise uncertainty) may lead to the phenomenon of the SNR wall, which is a SNR level below which the energy detector can not reliably detect any transmitted signal even increasing the observation time. It has also been shown that the SNR wall is not caused by the presence of a noise uncertainty itself, but by an insufficient refinement of the noise power estimation while the observation time increases.

- Cyclostationary-feature detection: These type of spectrum sensing algorithms are motivated because most man-made communication signals, such as BPSK, QPSK, AM, OFDM, etc. exhibit cyclostationary behavior. However, noise signals (typically white noise) do not exhibit cyclostationary behavior. These detectors are robust against noise variance uncertainty. The aim of such detectors is to exploit the cyclostationary nature of man-made communication signals buried in noise. Cyclostationary detectors can be either single cycle or multicycle cyclostatonary.

- *Wideband spectrum sensing*: refers to spectrum sensing over large spectral bandwidth, typically hundreds of MHz or even several GHz. Since current ADC technology cannot afford the high sampling rate with high resolution, it requires revolutional techniques, e.g., compressive sensing and sub-Nyquist sampling.

  o *Cooperative detection*: Refers to spectrum-sensing methods where information from multiple cognitive-radio users is incorporated for primary-user detection.

  o *Interference-based detection*.

- *Null-space based CR*: With the aid of multiple antennas, CR detects the null-space of the primary-user and then transmits within the null-space, such that its subsequent transmission causes less interference to the primary-user.

- *Spectrum management*: Capturing the best available spectrum to meet user communication requirements, while not creating undue interference to other (primary) users. Cognitive radios should decide on the best spectrum band (of all bands available) to meet quality of service requirements; therefore, spectrum-management functions are required for cognitive radios. Spectrum-management functions are classified as:

  o *Spectrum analysis*

  o *Spectrum decision*

The practical implementation of spectrum-management functions is a complex and multifaceted issue, since it must address a variety of technical and legal requirements. An example of the former is choosing an appropriate sensing threshold to detect other users, while the latter is exemplified by the need to meet the rules and regulations set out for radio spectrum access in international (ITU radio regulations) and national (telecommunications law) legislation.

Versus Intelligent Antenna (IA)

An intelligent antenna (or smart antenna) is an antenna technology that uses spatial beam-formation and spatial coding to cancel interference; however, applications are emerging for extension to intelligent multiple or cooperative-antenna arrays for application to complex communication environments. Cognitive radio, by comparison, allows user terminals to sense whether a portion of the spectrum is being used in order to share spectrum with neighbor users. The following table compares the two:

Point	Cognitive radio (CR)	Intelligent antenna (IA)
Principal goal	Open spectrum sharing	Ambient spatial reuse
Interference processing	Avoidance by spectrum sensing	Cancellation by spatial precoding/post-coding
Key cost	Spectrum sensing and multi-band RF	Multiple- or cooperative-antenna arrays
Challenging algorithm	Spectrum management tech	Intelligent spatial beamforming/coding tech
Applied techniques	Cognitive software radio	Generalized dirty paper coding and Wyner-Ziv coding
Basement approach	Orthogonal modulation	Cellular based smaller cell
Competitive technology	Ultra-wideband for greater band utilization	Multi-sectoring (3, 6, 9, so on) for higher spatial reuse
Summary	Cognitive spectrum-sharing technology	Intelligent spectrum reuse technology

Note that both techniques can be combined as illustrated in many contemporary transmission scenarios.

Cooperative MIMO (CO-MIMO) combines both techniques.

Applications

CR can sense its environment and, without the intervention of the user, can adapt to the user's communications needs while conforming to FCC rules in the United States. In theory, the amount of spectrum is infinite; practically, for propagation and other reasons it is finite because of the desirability of certain spectrum portions. Assigned spectrum is far from being fully utilized, and efficient spectrum use is a growing concern; CR offers a solution to this problem. A CR can intelligently detect whether any portion of the spectrum is in use, and can temporarily use it without interfering with the transmissions of other users. According to Bruce Fette, "Some of the radio's other cognitive abilities include determining its location, sensing spectrum use by neighboring devices, changing frequency, adjusting output power or even altering transmission parameters and characteristics. All of these capabilities, and others yet to be realized, will provide wireless spectrum users with the ability to adapt to real-time spectrum conditions, offering regulators, licenses and the general public flexible, efficient and comprehensive use of the spectrum".

Examples of applications include:

- The application of CR networks to emergency and public safety communications by utilizing white space

- The potential of CR networks for executing dynamic spectrum access (DSA)

- Application of CR networks to military action such as chemical biological radiological and nuclear attack detection and investigation, command control, obtaining information of battle damage evaluations, battlefield surveillance, intelligence assistance, and targeting.

Simulation of CR Networks

At present, modeling & simulation is the only paradigm which allows the simulation of complex behavior in a given environment's cognitive radio networks. Network simulators like OPNET, Net-Sim, MATLAB and NS2 can be used to simulate a cognitive radio network. Areas of research using network simulators include:

1. Spectrum sensing & incumbent detection

2. Spectrum allocation

3. Measurement and modeling of spectrum usage

4. Efficiency of spectrum utilization

Future Plans

The success of the unlicensed band in accommodating a range of wireless devices and services has led the FCC to consider opening further bands for unlicensed use. In contrast, the licensed bands are underutilized due to static frequency allocation. Realizing that CR technology has the potential to exploit the inefficiently utilized licensed bands without causing interference to incumbent users, the FCC released a Notice of Proposed Rule Making which would allow unlicensed radios to operate in the TV-broadcast bands. The IEEE 802.22 working group, formed in November 2004, is tasked with defining the air-interface standard for wireless regional area networks (based on CR sensing) for the operation of unlicensed devices in the spectrum allocated to TV service. To comply with later FCC regulations on unlicensed utilization of TV spectrum, the IEEE 802.22 has defined interfaces to the mandatory TV White Space Database in order to avoid interference to incumbent services.

Wireless Mesh Network

A wireless mesh network (WMN) is a communications network made up of radio nodes organized in a mesh topology. It is also a form of wireless ad hoc network.

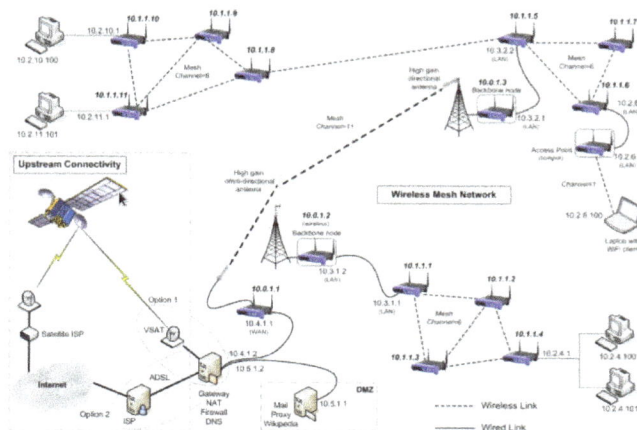

Diagram showing a possible configuration for a wired-wireless mesh network, connected upstream via a VSAT link (click to enlarge)

A mesh refers to rich interconnection among devices or nodes. Wireless mesh networks often consist of mesh clients, mesh routers and gateways. Mobility of nodes is less frequent. If nodes were to constantly or frequently move, the mesh will spend more time updating routes than delivering data. In a wireless mesh network, topology tends to be more static, so that routes computation can converge and delivery of data to their destinations can occur. Hence, this is a low-mobility centralized form of wireless ad hoc network. Also, because it sometimes relies on static nodes to act as gateways, it is not a truly all-wireless ad hoc network.

The mesh clients are often laptops, cell phones and other wireless devices while the mesh routers forward traffic to and from the gateways which may, but need not, be connected to the Internet. The coverage area of the radio nodes working as a single network is sometimes called a mesh cloud. Access to this mesh cloud is dependent on the radio nodes working in harmony with each other to create a radio network. A mesh network is reliable and offers redundancy. When one node can no longer operate, the rest of the nodes can still communicate with each other, directly or through one or more intermediate nodes. Wireless mesh networks can self form and self heal. Wireless mesh networks work with different wireless technologies including 802.11, 802.15, 802.16, cellular technologies and need not be restricted to any one technology or protocol.

History

Architecture

Wireless mesh architecture is a first step towards providing cost effective and low mobility over a specific coverage area. Wireless mesh infrastructure is, in effect, a network of routers minus the cabling between nodes. It's built of peer radio devices that don't have to be cabled to a wired port like traditional WLAN access points (AP) do. Mesh infrastructure carries data over large distances by splitting the distance into a series of short hops. Intermediate nodes not only boost the signal, but cooperatively pass data from point A to point B by making forwarding decisions based on their knowledge of the network, i.e. perform routing by first deriving the topology of the network.

Wireless mesh networks is a relatively "stable-topology" network except for the occasional failure of nodes or addition of new nodes. The path of traffic, being aggregated from a large number of end users, changes infrequently. Practically all the traffic in an infrastructure mesh network is either forwarded to or from a gateway, while in wireless ad hoc networks or client mesh networks the traffic flows between arbitrary pairs of nodes.

If rate of mobility among nodes are high, i.e., link breaks happen frequently, wireless mesh networks will start to break down and have low communication performance.

Management

This type of infrastructure can be decentralized (with no central server) or centrally managed (with a central server). Both are relatively inexpensive, and can be very reliable and resilient, as each node needs only transmit as far as the next node. Nodes act as routers to transmit data from nearby nodes to peers that are too far away to reach in a single hop, resulting in a network that can span larger distances. The topology of a mesh network has to be relatively stable, i.e., not too much mobility. If one node drops out of the network, due to hardware failure or any other reason, its neighbors can quickly find another route using a routing protocol.

Applications

Mesh networks may involve either fixed or mobile devices. The solutions are as diverse as communication needs, for example in difficult environments such as emergency situations, tunnels, oil rigs, battlefield surveillance, high-speed mobile-video applications on board public transport or real-time racing-car telemetry. An important possible application for wireless mesh networks is VoIP. By using a Quality of Service scheme, the wireless mesh may support local telephone calls to be routed through the mesh. Most applications in wireless mesh networks are similar to those in wireless ad hoc networks.

Some current applications:

- U.S. military forces are now using wireless mesh networking to connect their computers, mainly ruggedized laptops, in field operations.

- Electric smart meters now being deployed on residences transfer their readings from one to another and eventually to the central office for billing without the need for human meter readers or the need to connect the meters with cables.

- The laptops in the One Laptop per Child program use wireless mesh networking to enable students to exchange files and get on the Internet even though they lack wired or cell phone or other physical connections in their area.

- Google Home, Google Wi-Fi, and Google OnHub all support Wi-Fi mesh (i.e., Wi-Fi ad hoc) networking.

- The 66-satellite Iridium constellation operates as a mesh network, with wireless links between adjacent satellites. Calls between two satellite phones are routed through the mesh, from one satellite to another across the constellation, without having to go through an earth station. This makes for a smaller travel distance for the signal, reducing latency, and also allows for the constellation to operate with far fewer earth stations than would be required for 66 traditional communications satellites.

Operation

The principle is similar to the way packets travel around the wired Internet – data will hop from one device to another until it eventually reaches its destination. Dynamic routing algorithms implemented in each device allow this to happen. To implement such dynamic routing protocols, each device needs to communicate routing information to other devices in the network. Each device then determines what to do with the data it receives – either pass it on to the next device or keep it, depending on the protocol. The routing algorithm used should attempt to always ensure that the data takes the most appropriate (fastest) route to its destination.

Multi-radio Mesh

Multi-radio mesh refers to having different radios operating at different frequencies to interconnect nodes in a mesh. This means there is a unique frequency used for each wireless hop and thus a dedicated CSMA collision domain. With more radio bands, communication throughput is likely to increase as a result of more available communication channels. This is similar to providing dual or multiple radio paths to transmit and receive data.

Research Topics

One of the more often cited papers on Wireless Mesh Networks identified the following areas as open research problems in 2005

- New modulation scheme

    o In order to achieve higher transmission rate, new wideband transmission schemes other than OFDM and UWB are needed.

- Advanced antenna processing

    o Advanced antenna processing including directional, smart and multiple antenna technologies is further investigated, since their complexity and cost are still too high for wide commercialization.

- Flexible spectrum management

    o Tremendous efforts on research of frequency-agile techniques are being performed for increased efficiency.

- Cross-layer optimization

    o Cross-layer research is a popular current research topic where information is shared between different communications layers in order to increase the knowledge and current state of the network. This could enable new and more efficient protocols to be developed. A joint protocol which combines various design problems like routing, scheduling, channel assignment etc. can achieve higher performance since it is proven that these problems are strongly co-related. It is important to note that careless cross-layer design could lead to code which is difficult to maintain and extend.

- Software-defined wireless networking

  o Centralized, distributed, or hybrid? - In a new SDN architecture for WDNs is explored that eliminates the need for multi-hop flooding of route information and therefore enables WDNs to easily expand. The key idea is to split network control and data forwarding by using two separate frequency bands. The forwarding nodes and the SDN controller exchange link-state information and other network control signaling in one of the bands, while actual data forwarding takes place in the other band.

- Security

  o A WMN can be seen as a group of nodes (clients or routers) that cooperate to provide connectivity. Such an open architecture, where clients serve as routers to forward data packets, is exposed to many types of attacks that can interrupt the whole network and cause denial of service (DoS) or Distributed Denial of Service (DDoS).

Protocols

Routing Protocols

There are more than 70 competing schemes for routing packets across mesh networks. Some of these include:

- Associativity-Based Routing (ABR)

- AODV (Ad hoc On-Demand Distance Vector)

- B.A.T.M.A.N. (Better Approach To Mobile Adhoc Networking)

- Babel (protocol) (a distance-vector routing protocol for IPv6 and IPv4 with fast convergence properties)

- Dynamic NIx-Vector Routing|DNVR

- DSDV (Destination-Sequenced Distance-Vector Routing)

- DSR (Dynamic Source Routing)

- HSLS (Hazy-Sighted Link State)

- HWMP (Hybrid Wireless Mesh Protocol, the default mandatory routing protocol of IEEE 802.11s)

- *Infrastructure Wireless Mesh Protocol* (IWMP) for Infrastructure Mesh Networks by GRECO UFPB-Brazil

- OLSR (Optimized Link State Routing protocol)

- OORP (OrderOne Routing Protocol) (OrderOne Networks Routing Protocol)

- OSPF (Open Shortest Path First Routing)

- Routing Protocol for Low-Power and Lossy Networks (IETF ROLL RPL protocol, RFC 6550)

- PWRP (Predictive Wireless Routing Protocol)

- TORA (Temporally-Ordered Routing Algorithm)
- ZRP (Zone Routing Protocol)

The IEEE has developed a set of standards under the title 802.11s.

A less thorough list can be found at Ad hoc routing protocol list.

Autoconfiguration Protocols

Standard autoconfiguration protocols, such as DHCP or IPv6 stateless autoconfiguration may be used over mesh networks.

Mesh network specific autoconfiguration protocols include:

- Ad Hoc Configuration Protocol (AHCP)
- Proactive Autoconfiguration (Proactive Autoconfiguration Protocol)
- Dynamic WMN Configuration Protocol (DWCP)

Communities and Providers

- AWMN
- CUWiN
- Freifunk (DE) / FunkFeuer (AT) / OpenWireless (CH)
- Firechat
- Firetide
- Guifi.net
- Netsukuku
- Ninux (IT)

Products

- Aruba AirMesh- multiservice wireless mesh networks for outdoors
- Ruckus Mesh - Smart Mesh
- FireTide - Wireless mesh networks
- Cisco Meraki - Mesh networking - access points as gateways and repeaters
- Juniper Wireless Mesh - Wireless mesh and bridging
- OpenMesh - Open Mesh makes WiFi smarter and simpler
- StrixSystems - 802.11n powered wireless mesh

- MeshDynamics - Meshdynamics Industrial Mesh Networks

- Rajant Mesh - Kinetic Mesh Networks

- Others - list of venture backed mesh networking companies.

References

- S. Haykin, "Cognitive Radio: Brain-empowered Wireless Communications", IEEE Journal on Selected Areas of Communications, vol. 23, nr. 2, pp. 201–220, Feb. 2005

- Software Defined Radio: Architectures, Systems and Functions (Markus Dillinger, Kambiz Madani, Nancy Alonistioti) Page xxxiii (Wiley & Sons, 2003, ISBN 0-470-85164-3)

- Rick Lindquist; Joel R. Hailas (October 2005). "FlexRadio Systems; SDR-1000 HF+VHF Software Defined Radio Redux". QST. Retrieved 2008-12-07

- H. Urkowitz "Energy detection of unknown deterministic signals", IEEE Proceedings, Apr. 1967. doi:10.1109/PROC.1967.5573

- Ian F. Akyildiz, W.-Y. Lee, M. C. Vuran, and S. Mohanty, "NeXt Generation/Dynamic Spectrum Access/Cognitive Radio Wireless Networks: A Survey," Computer Networks (Elsevier) Journal, September 2006

- Mitola III, J. (1992). Software radios-survey, critical evaluation and future directions. National Telesystems Conference. pp. 13/15 to 13/23. ISBN 0-7803-0554-X. doi:10.1109/NTC.1992.267870

- R. Tandra and A. Sahai, "SNR walls for signal detection", IEEE J. Sel. Topics Signal Process., vol. 2, no. 1, pp. 4–17, Feb. 2008. doi:10.1109/JSTSP.2007.914879

- Carlos Cordeiro, Kiran Challapali, and Dagnachew Birru. Sai Shankar N. IEEE 802.22: An Introduction to the First Wireless Standard based on Cognitive Radios JOURNAL OF COMMUNICATIONS, VOL. 1, NO. 1, APRIL 2006

- W. A. Gardner, "Exploitation of spectral redundancy in cyclostationary signals", IEEE Sig. Proc. Mag., vol. 8, no. 2, pp. 14–36, 1991. doi:10.1109/79.81007

- Khattab, Ahmed; Perkins, Dmitri; Bayoumi, Magdy (2013-01-01). Cognitive Radio Networks. Analog Circuits and Signal Processing. Springer New York. pp. 33–39. ISBN 9781461440321. doi:10.1007/978-1-4614-4033-8_4

- Mitola III, Joseph. "Cognitive Radio An Integrated Agent Architecture for Software Defined Radio" (PDF). Archived from the original (PDF) on 17 September 2012. Retrieved 7 January 2013

- Tallon, J.; Forde, T. K.; Doyle, L. (2012-06-01). "Dynamic Spectrum Access Networks: Independent Coalition Formation". IEEE Vehicular Technology Magazine. 7 (2): 69–76. ISSN 1556-6072. doi:10.1109/MVT.2012.2190218

An Overview of Radio Frequency Engineering

Radio frequency can be measured in units called hertz. Radio frequency engineering is a field of study of electrical engineering which studies devices that are used to control radio frequency spectrum. This chapter is an overview of the subject matter incorporating all the major aspects of radio frequency.

Radio Frequency

Radio frequency (RF) is any of the electromagnetic wave frequencies that lie in the range extending from around 3 kHz to 300 GHz, which include those frequencies used for communications or radar signals. RF usually refers to electrical rather than mechanical oscillations. However, mechanical RF systems do exist.

Although radio *frequency* is a rate of oscillation, the term "radio frequency" or its abbreviation "RF" are used as a synonym for radio – i.e., to describe the use of wireless communication, as opposed to communication via electric wires. Examples include:

- Radio-frequency identification
- ISO/IEC 14443-2 *Radio frequency power and signal interface*

Special Properties of RF Current

Electric currents that oscillate at radio frequencies have special properties not shared by direct current or alternating current of lower frequencies.

- RF current does not penetrate deeply into electrical conductors but tends to flow along their surfaces; this is known as the skin effect.

- RF currents applied to the body are harmful, but often do not cause the painful sensation of electric shock that lower frequency currents produce. This is because the current changes direction too quickly to trigger depolarization of nerve membranes. However they can cause serious superficial burns called *RF burns*.

- RF current can easily ionize air, creating a conductive path through it. This property is exploited by "high frequency" units used in electric arc welding, which use currents at higher frequencies than power distribution uses.

- Another property is the ability to appear to flow through paths that contain insulating material, like the dielectric insulator of a capacitor. This is because capacitive reactance in a circuit decreases with frequency.

- In contrast, RF current can be blocked by a coil of wire, or even a single turn or bend in a wire. This is because the inductive reactance of a circuit increases with frequency.

- When conducted by an ordinary electric cable, RF current has a tendency to reflect from discontinuities in the cable such as connectors and travel back down the cable toward the source, causing a condition called standing waves. Therefore, RF current must be carried by specialized types of cable called transmission line.

Radio Communication

To receive radio signals an antenna must be used. However, since the antenna will pick up thousands of radio signals at a time, a radio tuner is necessary to *tune into* a particular frequency (or frequency range). This is typically done via a resonator – in its simplest form, a circuit with a capacitor and an inductor form a tuned circuit. The resonator amplifies oscillations within a particular frequency band, while reducing oscillations at other frequencies outside the band. Another method to isolate a particular radio frequency is by oversampling (which gets a wide range of frequencies) and picking out the frequencies of interest, as done in software defined radio.

The distance over which radio communications is useful depends significantly on things other than wavelength, such as transmitter power, receiver quality, type, size, and height of antenna, mode of transmission, noise, and interfering signals. Ground waves, tropospheric scatter and skywaves can all achieve greater ranges than line-of-sight propagation. The study of radio propagation allows estimates of useful range to be made.

Frequency Bands

Frequency	Wavelength	Designation	Abbreviation
3–30 Hz	10^5–10^4 km	Extremely low frequency	ELF
30–300 Hz	10^4–10^3 km	Super low frequency	SLF
300–3000 Hz	10^3–100 km	Ultra low frequency	ULF
3–30 kHz	100–10 km	Very low frequency	VLF
30–300 kHz	10–1 km	Low frequency	LF
300 kHz – 3 MHz	1 km – 100 m	Medium frequency	MF
3–30 MHz	100–10 m	High frequency	HF
30–300 MHz	10–1 m	Very high frequency	VHF
300 MHz – 3 GHz	1 m – 10 cm	Ultra high frequency	UHF
3–30 GHz	10–1 cm	Super high frequency	SHF
30–300 GHz	1 cm – 1 mm	Extremely high frequency	EHF
300 GHz – 3 THz	1 mm – 0.1 mm	Tremendously high frequency	THF

In Medicine

Radio frequency (RF) energy, in the form of radiating waves or electrical currents, has been used in medical treatments for over 75 years, generally for minimally invasive surgeries using radiofrequency ablation including the treatment of sleep apnea. Magnetic resonance imaging (MRI) uses radio frequency waves to generate images of the human body.

Radio frequencies at non-ablation energy levels are sometimes used as a form of cosmetic treatment that can tighten skin, reduce fat (lipolysis), or promote healing.

RF diathermy is a medical treatment that uses RF induced heat as a form of physical or occupational therapy and in surgical procedures. It is commonly used for muscle relaxation. It is also a method of heating tissue electromagnetically for therapeutic purposes in medicine. Diathermy is used in physical therapy and occupational therapy to deliver moderate heat directly to pathologic lesions in the deeper tissues of the body. Surgically, the extreme heat that can be produced by diathermy may be used to destroy neoplasms, warts, and infected tissues, and to cauterize blood vessels to prevent excessive bleeding. The technique is particularly valuable in neurosurgery and surgery of the eye. Diathermy equipment typically operates in the short-wave radio frequency (range 1–100 MHz) or microwave energy (range 434–915 MHz).

Pulsed electromagnetic field therapy (PEMF) is a medical treatment that purportedly helps to heal bone tissue reported in a recent NASA study. This method usually employs electromagnetic radiation of different frequencies - ranging from static magnetic fields, through extremely low frequencies (ELF) to higher radio frequencies (RF) administered in pulses.

Effects on the Human Body

Extremely Low Frequency RF

High-power extremely low frequency RF with electric field levels in the low kV/m range are known to induce perceivable currents within the human body that create an annoying tingling sensation. These currents will typically flow to ground through a body contact surface such as the feet, or arc to ground where the body is well insulated.

Microwaves

Microwave exposure at low-power levels below the Specific absorption rate set by government regulatory bodies are considered harmless non-ionizing radiation and have no effect on the human body. However, levels above the Specific absorption rate set by the U.S. Federal Communications Commission are considered potentially harmful.

Long-term human exposure to high-levels of microwaves is recognized to cause cataracts according to experimental animal studies and epidemiological studies. The mechanism is unclear but may include changes in heat sensitive enzymes that normally protect cell proteins in the lens. Another mechanism that has been advanced is direct damage to the lens from pressure waves induced in the aqueous humor.

High-power exposure to microwave RF is known to create a range of effects from lower to higher power levels, ranging from unpleasant burning sensation on the skin and microwave auditory effect, to extreme pain at the mid-range, to physical burning and blistering of skin and internals at high power levels.

General RF Exposure

The 1999 revision of Canadian Safety Code 6 recommended electric field limits of 100 kV/m for pulsed EMF to prevent air breakdown and spark discharges, mentioning rationale related to auditory effect and energy-induced unconsciousness in rats. The pulsed EMF limit was removed in later revisions, however.

As a Weapon

A heat ray is an RF harassment device that makes use of microwave radio frequencies to create an unpleasant heating effect in the upper layer of the skin. A publicly known heat ray weapon called the Active Denial System was developed by the US military as an experimental weapon to deny the enemy access to an area. A death ray is a weapon that delivers heat ray electromagnetic energy at levels that injure human tissue. The inventor of the death ray, Harry Grindell Matthews, claims to have lost sight in his left eye while developing his death ray weapon based on a primitive microwave magnetron from the 1920s (note that a typical microwave oven induces a tissue damaging cooking effect inside the oven at about 2 kV/m.).

Measurement

Since radio frequency radiation has both an electric and a magnetic component, it is often convenient to express intensity of radiation field in terms of units specific to each component. The unit *volts per meter* (V/m) is used for the electric component, and the unit *amperes per meter* (A/m) is used for the magnetic component. One can speak of an electromagnetic field, and these units are used to provide information about the levels of electric and magnetic field strength at a measurement location.

Another commonly used unit for characterizing an RF electromagnetic field is *power density*. Power density is most accurately used when the point of measurement is far enough away from the RF emitter to be located in what is referred to as the far field zone of the radiation pattern. In closer proximity to the transmitter, i.e., in the "near field" zone, the physical relationships between the electric and magnetic components of the field can be complex, and it is best to use the field strength units discussed above. Power density is measured in terms of power per unit area, for example, milliwatts per square centimeter (mW/cm^2). When speaking of frequencies in the microwave range and higher, power density is usually used to express intensity since exposures that might occur would likely be in the far field zone.

Radio-frequency Engineering

Radio-frequency engineering is a subset of electrical engineering that deals with devices that are designed to operate in the radio frequency (RF) spectrum. These devices operate within the range of about 3 kHz up to 300 GHz.

Radio-frequency engineering is incorporated into almost everything that transmits or receives a radio wave, which includes, but is not limited to, mobile phones, radios, Wi-Fi, and two-way radios.

Radio-frequency engineering is a highly specialized field falling typically in one of two areas:

1. providing or controlling coverage with some kind of antenna/transmission system

2. generating or receiving signals to or from that transmission system to other communications electronics or controls.

To produce quality results, an in-depth knowledge of mathematics, physics, general electronics theory as well as specialized training in areas such as wave propagation, impedance transformations, filters, microstrip circuit board design, etc. may be required. Because of the many ways RF is conducted both through typical conductors as well as through space, an initial design of an RF circuit usually bears very little resemblance to the final optimized physical circuit. Revisions to the design are often required to achieve intended results.

Radio Electronics

Radio electronics is concerned with electronic circuits which receive or transmit radio signals.

Typically, such circuits must operate at radio frequency and power levels, which imposes special constraints on their design. These constraints increase in their importance with higher frequencies. At microwave frequencies, the reactance of signal traces becomes a crucial part of the physical layout of the circuit.

List of radio electronics topics:

- RF oscillators: PLL, Voltage-controlled oscillator
- Transmitters, transmission lines, RF connectors
- Antennas, antenna theory, list of antenna terms
- Receivers, tuners
- Amplifiers
- Modulators, demodulators, detectors
- RF filters
- RF shielding, Ground plane
- PCB layout guidelines
- DSSS, noise power
- Digital radio

Duties

Radio-frequency engineers are specialists in their respective field and can take on many different roles, such as design, installation, and maintenance. Radio-frequency engineers require many years of extensive experience in the area of study. This type of engineer has experience with transmission systems, device design, and placement of antennas for optimum performance. A radio-frequency engineer at a broadcast facility is responsible for maintenance of the stations high-power broadcast transmitters and associated systems. This includes transmitter site emergency power, remote control, main transmission line and antenna adjustments, microwave radio relay STL/TSL links, and more.

In addition, a radio-frequency design engineer must be able to understand electronic hardware design, circuit board material, antenna radiation, and the effect of interfering frequencies that prevent optimum performance within the piece of equipment being developed.

RF Connector

A type N coaxial RF connector (male)

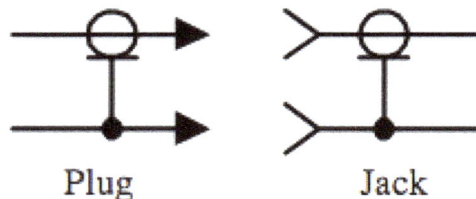

Plug   Jack

Electronic symbols for the plug and jack coaxial connectors

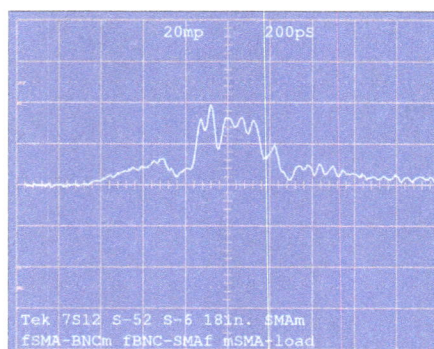

Time-domain reflectometry shows reflections due to impedance variations in mated RF connectors.

A coaxial RF connector (radio frequency connector) is an electrical connector designed to work at radio frequencies in the multi-megahertz range. RF connectors are typically used with coaxial cables and are designed to maintain the shielding that the coaxial design offers. Better models also minimize the change in transmission line impedance at the connection. Mechanically, they may provide a fastening mechanism (thread, bayonet, braces, blind mate) and springs for a low ohmic electric contact while sparing the gold surface, thus allowing very high mating cycles and reducing the insertion force. Research activity in the area of radio-frequency (RF) circuit design has surged in the 2000s in direct response to the enormous market demand for inexpensive, high-data-rate wireless transceivers.

Tuner (Radio)

Opened VHF/UHF tuner of a television set. The antenna connector is on the right.

A tuner is a subsystem that receives radio frequency (RF) transmissions like radio broadcasts and converts the selected carrier frequency and its associated bandwidth into a fixed frequency that is suitable for further processing, usually because a lower frequency is used on the output. Broadcast FM/AM transmissions usually feed this intermediate frequency (IF) directly into a demodulator that convert the radio signal into audio-frequency signals that can be fed into an amplifier to drive a loudspeaker.

More complex transmissions like PAL/NTSC (TV), DAB (digital radio), DVB-T/DVB-S/DVB-C (digital TV) etc. use a wider frequency bandwidth, often with several subcarriers. These are transmitted inside the receiver as an intermediate frequency (IF). The next step is usually either to process subcarriers like real radio transmissions or to sample the whole bandwidth with A/D at a rate faster than the Nyquist rate that is at least 2 times the IF frequency.

The tuner can also refer to a radio receiver or standalone audio component that are part of an audio system, to be connected to a separate amplifier. The verb tuning in radio contexts means adjusting the radio receiver to receive the desired radio signal carrier frequency that a particular radio station uses.

Design

The simplest tuner consists of an inductor and capacitor connected in parallel, where the capacitor or inductor is made to be variable. This creates a resonant circuit which responds to an alternating current at one frequency. Combined with a detector, also known as a demodulator (diode D1 in the circuit), it becomes the simplest radio receiver, often called a crystal set.

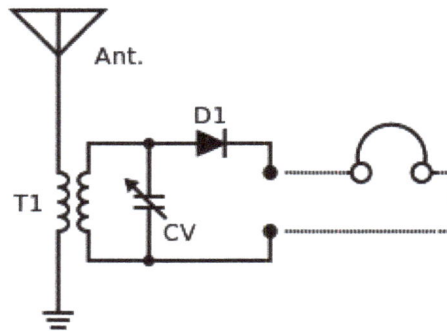

Inductively coupled crystal radio receiver

Older models would realize manual tuning by means of mechanically operated ganged variable capacitors. Often several sections would be provided on a tuning capacitor, to tune several stages of the receiver in tandem, or to allow switching between different frequency bands. A later method used a potentiometer supplying a variable voltage to varactor diodes in the local oscillator and tank circuits of front end tuner, for electronic tuning.

Modern radio tuners use a superheterodyne receiver with tuning selected by adjustment of the frequency of a local oscillator. This system shifts the radio frequency of interest to a fixed frequency so that it can be tuned with fixed-frequency band-pass filter. Still later, phase locked loop methods were used, with microprocessor control.

In a self-contained radio receiver for audio, the signal from the detector after the tuner is run through a volume control and to an amplifier stage. The amplifier feeds either an internal speaker or headphones. In a tuner component of an audio system (for example, a home high-fidelity system or a public address system in a building), the output of the detector is connected to a separate external system of amplifiers and speakers.

The broadcast audio FM band (88 - 108 MHz in most countries) is around 100 times higher in frequency than the AM band and provides enough space for a bandwidth of 50 kHz. This bandwidth is sufficient to transmit both stereo channels with almost the full hearing range. Sometimes, additional subcarriers are used for unrelated audio or data transmissions. The left and right audio signals must be combined into a single signal which is applied to the modulation input of the transmitter; this is done by the addition of an inaudible subcarrier signal to the FM broadcast signal. FM stereo allows left and right channels to be transmitted. The availability of FM stereo, a quieter VHF broadcast band, and better fidelity led to the specialization of FM broadcasting in music, tending to leave AM broadcasting with spoken-word material.

Restoration

Standalone audio stereo FM tuners are sought after for audiophile and TV/FM DX applications, especially those produced in the 1970s and early 1980s, when performance and manufacturing standards were among the highest. In many instances the tuner may be modified to improve performance. A growing hobby trend is the electronics specialists that buy, collect and restore these vintage FM or AM/FM audio tuners. The restoration usually begins with replacing the electrolytic capacitors that may age over time. The tuner is outfitted with improved tolerance and better sounding upgraded parts. Prices have increased relative to the increasing demand for the older audio tuners. Those with the most value are the best sounding, most rare (collectible), the best DX capable (distance reception) and the known build quality of the component, as it left the factory.

AM/FM

Most of the early tuner models were designed and manufactured to receive only the AM broadcast band. As FM became more popular, the limitations of AM became more apparent, and FM became the primary listening focus, especially for stereo and music broadcasting. Few companies even manufacture dedicated FM or AM/FM tuners now, as these bands are most often included in a low cost chip for A/V systems, more as an afterthought, rather than designed for the critical FM listener.

In Europe, where a second AM broadcast band is used for longwave broadcasting, tuners may be fitted with both the standard medium wave and the additional longwave band. However, radios with only medium wave are also common, especially in countries where there are no longwave broadcasters. Rarely, radios are sold with only FM and longwave, but no medium wave band. Some tuners may also be equipped with one or more short wave bands.

Television

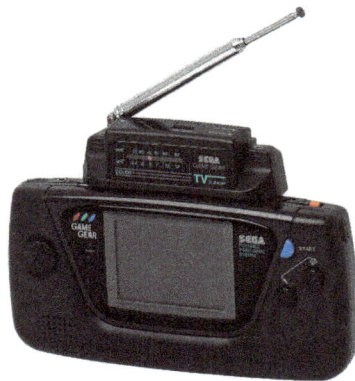

A TV Tuner plugged into Sega Game Gear.

A *television tuner* converts a radio frequency analog television or digital television transmission into audio and video signals which can be further processed to produce sound and a picture. Different tuners are used for different television standards such as PAL, NTSC, ATSC, SECAM, DVB-C, DVB-T, DVB-T2, ISDB, T-DMB, open cable. An example frequency range is 48.25 MHz - 855.25 MHz (E2-E69), with a tuning frequency step size of 31.25, 50 or 62.5 kHz. Modern solid-state internal TV-tuner modules typically weigh around 45 g.

Analog tuners can tune only analog signals. An ATSC tuner is a digital tuner that tunes digital signals only. Some digital tuners provide an analog bypass.

VHF/UHF TV tuners are rarely found as a separate component, but are incorporated into television sets. Cable boxes and other set top boxes contain tuners for digital TV services, and send their output via SCART or other connector, or using an RF modulator (typically on channel 36 in Europe and channel 3/4 in North America) to TV receivers that do not natively support the services.

They provide outputs via composite, S-video, or component video. Many can be used with video monitors that do not have a TV tuner or direct video input. They are often part of a VCR or digital video recorder (DVR, PVR). Many home computers in the 1970s and 1980s used an RF modulator to connect to a TV set.

Personal computers may be fitted with expansion cards (typically with PCI or USB interface) providing a TV tuner and digital signal processor (DSP). They may be dedicated TV tuner cards, or incorporated into a video card. These cards allow a computer to display and capture television programs. Many earlier models were stand-alone tuners, designed only to deliver TV pictures through a VGA connector; this allowed viewing television on a computer display, but did not support recording television programs.

Smartphone and tablets can use a Micro USB DVB-T receiver to watch DVB-T TV.

Electronic Tuner

An electronic tuner is a device which tunes across a part of the radio frequency spectrum by the application of a voltage or appropriate digital code words. This type of tuner supersedes mechanical tuners, which were tuned by manual adjustment of capacitance or inductance in the tuned circuits. In a more practical and everyday sense, a radio or television set which is tuned by manually turning a knob or dial contains a manual tuner into which the shaft of that knob or dial extends.

Early model televisions and radios were tuned by a rack of momentary push buttons; some of the earlier types were purely mechanical and adjusted the capacitance or inductance of the tuned circuit to a preset number of positions corresponding to the frequencies of popular local stations. Later electronic types utilized the varactor diode as a voltage controlled capacitance in the tuned circuit, to receive a number of preset voltages from the rack of buttons tuning the device instantly to local stations. The mechanical button rack was popular in car radios of the 1960s and 1970s. The electronic button rack controlling the new electronic varactor tuner was popular in television sets of the 1970s and 1980s.

Modern electronic tuners also use varactor diodes as the actual tuning elements, but the voltages which change their capacitance are obtained from a digital to analog converter (DAC) driven by a microprocessor or phase locked loop (PLL) arrangement. This modern form allows for very precise tuning and locking-in on weak signals, as well as a numerical display of the tuned frequency.

References

- Ruey J. Sung & Michael R. Lauer (2000). Fundamental approaches to the management of cardiac arrhythmias. Springer. p. 153. ISBN 978-0-7923-6559-4

- "ISO/IEC 14443-2:2001 Identification cards — Contactless integrated circuit(s) cards — Proximity cards — Part 2: Radio frequency power and signal interface". Iso.org. 2010-08-19. Retrieved 2011-11-08

- Melvin A. Shiffman; Sid J. Mirrafati; Samuel M. Lam; Chelso G. Cueteaux (2007). Simplified Facial Rejuvenation. Springer. p. 157. ISBN 978-3-540-71096-7

- "Definition of RADIO FREQUENCY". Merriam-Webster. Encyclopædia Britannica. n.d. Retrieved 6 August 2015

- Jeffrey S. Beasley; Gary M. Miller (2008). Modern Electronic Communication (9th ed.). pp. 4–5. ISBN 978-0132251136

Various Radio Electronics

Radio receiver, radio transmitter design, radio repeater, absorption wavemeter, crystal radio and lecher lines are the various kinds of radio electronics. Radio receivers are devices used to convert radio waves into information which can be carried forward. This chapter has been carefully written to provide an easy understanding of the various radio electronics.

Radio Receiver

Early broadcast radio receiver. Truetone model from about 1940

In radio communications, a radio receiver (radio) is an electronic device that receives radio waves and converts the information carried by them to a usable form. It is used with an antenna. The antenna intercepts radio waves (electromagnetic waves) and converts them to tiny alternating currents which are applied to the receiver, and the receiver extracts the desired information. The receiver uses electronic filters to separate the desired radio frequency signal from all the other signals picked up by the antenna, an electronic amplifier to increase the power of the signal for further processing, and finally recovers the desired information through demodulation.

The information produced by the receiver may be in the form of sound, images, or data. A radio receiver may be a separate piece of electronic equipment, or an electronic circuit within another device. Radio receivers are components of communications, broadcasting, remote control, and network systems. In consumer electronics, the terms *radio* and *radio receiver* are often used specifically for receivers designed to reproduce sound transmitted by radio broadcasting stations, historically the first mass-market commercial radio application.

Broadcast Radio Receivers

The most familiar form of radio receiver is a broadcast receiver, often just called a *radio*, which receives audio programs intended for public reception transmitted by local radio stations. The sound is reproduced either by a loudspeaker in the radio or an earphone which plugs into a jack on the radio. The radio requires electric power, provided either by batteries inside the radio or a power cord which plugs into an electric outlet. All radios have a volume control to adjust the loudness of the audio, and some type of "tuning" control to select the radio station to be received.

Tuning

Radio waves transmitted by multiple radio stations travel through the air simultaneously without interfering with each other because they have different frequencies; their carrier waves oscillate at different rates, measured in kilohertz (kHz) or megahertz (MHz). The frequency of radio stations is usually listed prominently in their advertising. In order to select a particular station to receive, the radio is adjusted ("*tuned*") to the frequency of the desired transmitter. In some radios this is done by the user turning a "tuning" knob until the desired station is heard in the radio's loudspeaker. In newer radios this is often done by pressing a "channel up" or "channel down" button, which causes the radio to automatically receive the next radio station it finds going up or down the frequency band. The radio has a dial or LCD display showing the frequency it is tuned to.

AM and FM

Modulation is the process of adding information to a radio carrier wave. In Amplitude modulation (AM) the strength of the radio signal is varied by the audio signal. The AM broadcast bands are between 148 and 283 kHz in the longwave range, and between 526 and 1706 kHz in the medium frequency (MF) range of the radio spectrum. AM broadcasting is also carried out in shortwave bands, between about 2.3 and 26 MHz.

In frequency modulation (FM) the frequency of the radio signal is varied slightly by the audio signal. FM broadcasting is permitted in the FM broadcast bands between about 65 and 108 MHz in the very high frequency (VHF) range. The exact frequency ranges vary somewhat in different countries. "AM/FM" radios have a switch to select which band to receive.

FM stereo radio stations broadcast in stereophonic sound (stereo), transmitting two sound channels representing left and right microphones. A stereo receiver contains the additional circuits and parallel signal paths to reproduce the two separate channels. While AM stereo transmitters and receivers exist, they have not achieved the popularity of FM stereo.

Reception

The signal strength of radio waves decreases the farther they travel from the transmitter, so the range of a system depends on the power of the transmitter and the sensitivity of the receiver. AM broadcast band radio waves travel as ground waves which follow the contour of the Earth, so AM radio stations can be reliably received at hundreds of miles distance. Due to their higher frequency, FM band radio signals do not travel far beyond the visual horizon; limiting reception distance to about 40 miles (64 km), and can be blocked by hills between the transmitter and receiver. However FM radio is less susceptible to interference from radio noise (RFI, sferics, static) and has higher fidelity; better frequency response and less audio distortion, than AM. So in many countries serious music is only broadcast by FM stations, and AM stations specialize in radio news, talk radio, and sports.

Types of Broadcast Receiver

Radios are made in a range of styles

- *Table radio* - A self-contained radio with speaker designed to sit on a table.

- *Clock radio* - A bedside table radio that also includes an alarm clock. The alarm clock can be set to turn on the radio in the morning instead of an alarm, to wake the owner.

- *Tuner* - A high fidelity AM/FM radio receiver in a component home audio system. It has no speakers but outputs an audio signal which is fed into the system and played through the system's speakers.

- *Portable radio* - a radio powered by batteries that can be carried with a person. Radios are now often integrated with other audio sources in CD players and portable media players.

  o *Boom box* - a portable battery-powered high fidelity stereo sound system in the form of a box with a handle.

  o *Transistor radio* - an older term for a portable pocket-sized broadcast radio receiver. Made possible by the invention of the transistor and developed in the 1960s, transistor radios were hugely popular during the 1960s and 70s, and changed the public's listening habits.

- *Car radio* - An AM/FM radio integrated into the dashboard of a vehicle, used for entertainment while driving. Virtually all cars and trucks come with car radios. It usually also includes a CD player.

- *Satellite radio* receiver - subscription radio receiver that receives audio programming from a direct broadcast satellite. The subscriber must pay a monthly fee. They are mostly designed as car radios.

- *Shortwave radio* - This is a broadcast radio that also receives the shortwave bands. It is used for shortwave listening.

- AV receivers are a common component in a high-fidelity or home-theatre system; in addition to receiving radio programming, the receiver will also contain switching and amplifying functions to interconnect and control the other components of the system.

Other Applications

Radio receivers are essential components of all systems that use radio. Besides broadcast receivers, described above, radio receivers are used in a huge variety of electronic systems in modern technology. They can be a separate piece of equipment (a *radio*), or a subsystem incorporated into other electronic devices. A transceiver is a transmitter and receiver combined in one unit. Below is a list of a few of the most common types, organized by function.

- Broadcast television reception - Televisions receive a video signal representing a moving image, composed of a sequence of still images, and a synchronized audio signal representing the associated sound. The television channel received by a TV occupies a wider bandwidth than an audio signal, from 600 kHz to 6 MHz.

    o *Terrestrial television receiver, broadcast television* or just *television* (TV) - Televisions contains an integral receiver (TV tuner) which receives free broadcast television from local television stations on TV channels in the VHF and UHF bands.

    o *Satellite TV* receiver - a set-top box which receives subscription direct-broadcast satellite television, and displays it on an ordinary television. A rooftop satellite dish receives many channels all modulated on a K_u band microwave downlink signal from a geostationary direct broadcast satellite 22,000 miles above the Earth, and the signal is converted to a lower intermediate frequency and transported to the box through a coaxial cable. The subscriber pays a monthly fee.

- Two-way voice communications - A two-way radio is an audio transceiver, a receiver and transmitter in the same device, used for bidirectional person-to-person voice communication. The radio link may be half-duplex, using a single radio channel in which only one radio can transmit at a time. so different users take turns talking, pressing a push to talk button on their radio which switches on the transmitter. Or the radio link may be full duplex, a bidirectional link using two radio channels so both people can talk at the same time, as in a cell phone.

    o *Cellphone* - a portable telephone that is connected to the telephone network by radio signals exchanged with a local antenna called a cell tower. Cellphones have highly automated digital receivers working in the UHF and microwave band that receive the incoming side of the duplex voice channel, as well as a control channel that handles dialing calls and switching the phone between cell towers. They usually also have several other receivers that connect them with other networks: a WiFi modem, a bluetooth modem, and a GPS receiver. The cell tower has sophisticated multichannel receivers that receive the signals from many cell phones simultaneously.

    o *Cordless phone* - a landline telephone in which the handset is portable and communicates with the rest of the phone by a short range duplex radio link, instead of being attached by a cord. Both the handset and the base station have radio receivers operating in the UHF band that receive the short range bidirectional duplex radio link.

    o *Citizens band radio* - a limited power half-duplex two-way radio operating in the 27 MHz band that can be used without a license. They are often installed in vehicles and used by truckers and delivery services.

    o *Walkie-talkie* - a handheld short range half-duplex two-way radio that communicates on VHF or UHF bands with other walkie-talkies.

Handheld scanner

Scanner - a receiver that continuously monitors multiple frequencies or radio channels by stepping through the channels repeatedly, listening briefly to each channel for a transmission. When a transmitter is found the receiver stops at that channel. Scanners are used to monitor emergency police, fire, and ambulance frequencies, as well as other two way radio frequencies such as citizens band. Scanning capabilities have also become a standard feature in communications receivers, walkie-talkies, and other two-way radios.

Modern communications receiver, ICOM RC-9500

Communications receiver or *shortwave receiver* - a general purpose audio receiver covering the LF, MF, shortwave (HF), and VHF bands. Used mostly with a separate shortwave transmitter for two-way voice communication in communication stations, amateur radio stations, and for shortwave listening.

- One-way (simplex) voice communications

    o *Wireless microphone* receiver - these receive the short range signal from wireless microphones used onstage by musical artists, public speakers, and television personalities.

Baby monitor. The receiver is on the left

Baby monitor - this is a cribside appliance for mothers of infants that transmits the baby's sounds to a receiver carried by the mother, so she can monitor the baby while she is in other parts of the house. Many baby monitors now have video cameras to show a picture of the baby.

- Data communications

  o *Wireless (WiFi) modem* - an automated short range digital data transmitter and receiver on a portable wireless device that communicates by microwaves with a nearby access point, a router or gateway, connecting the portable device with a local computer network (WLAN) to exchange data with other devices.

  o *Bluetooth* modem - a very short range (up to 10 m) 2.4-2.83 GHz data transceiver on a portable wireless device used as a substitute for a wire or cable connection, mainly to exchange files between portable devices and connect cellphones and music players with wireless earphones.

  o *Microwave relay* - a long distance high bandwidth point-to-point data transmission link consisting of a dish antenna and transmitter that transmits a beam of microwaves to another dish antenna and receiver. Since the antennas must be in line-of-sight, distances are limited by the visual horizon to 30-40 miles. Microwave links are used for private business data, wide area computer networks (WANs), and by telephone companies to transmit distance phone calls and television signals between cities.

- Satellite communications - Communication satellites are used for data transmission between widely separated points on Earth. Other satellites are used for search and rescue, remote sensing, weather reporting and scientific research. Radio communication with satellites and spacecraft can involve very long path lengths, from 35,786 km (22,236 mi) for geosynchronous satellites to billions of kilometers for interplanetary spacecraft. This and the limited power available to a spacecraft transmitter mean very sensitive receivers must be used.

  o *Satellite transponder* - A receiver and transmitter in a communications satellite that receives multiple data channels carrying long distance telephone calls, television signals. or internet traffic on a microwave uplink signal from a satellite ground station and retransmits the data to another ground station on a different downlink frequency. In a direct broadcast satellite the transponder broadcasts a stronger signal directly to satellite radio or satellite television receivers in consumer's homes.

  o *Satellite ground station receiver* - communication satellite ground stations receive data from communications satellites orbiting the Earth. Deep space ground stations such as those of the NASA Deep Space Network receive the weak signals from distant scientific spacecraft on interplanetary exploration missions. These have large dish antennas around 85 ft (25 m) in diameter, and extremely sensitive radio receivers similar to radio telescopes. The RF front end of the receiver is often cryogenically cooled to −195.79 °C (−320 °F) by liquid nitrogen to reduce radio noise in the circuit.

- Remote control - Remote control receivers are automated receivers in autonomous (unmanned) devices such as drone vehicles that receive digital commands that control the device. Remote control systems often also incorporate a telemetry channel to transmit data on the state of the controlled device back to the controller.

- o *Drone* receiver - unmanned drone aircraft are controlled by encrypted commands from a satellite.

- o *Radio controlled model* receiver - A receiver used to receive commands transmitted from a nearby handheld controller to steer radio-controlled toy cars, boats, airplanes, and helicopters. Usually operates on the 2.4 GHz band.

- o *Garage door opener* - Residential garage doors are opened and closed by the owner pressing a button on a transmitter in his car. The device transmits a short-range 2.45 GHz coded signal to a receiver in the garage door opener. The digital signal is encrypted and changes with each use, to prevent thieves from recording and copying it.

- o *Keyless lock* - most modern cars have keyless entry systems. To unlock the door, instead of inserting a key in a lock, the owner presses a button on a small transmitter on a keyring, which transmits a short range 2.45 GHz coded signal to a receiver in the car.

- Radiolocation - This is the use of radio waves to determine the location or direction of an object.

  - o *Radar* - a device that transmits a narrow beam of microwaves which reflect from a target back to a receiver, used to locate objects such as aircraft, spacecraft, missiles, ships or land vehicles. The reflected waves from the target are received by a receiver usually connected to the same antenna, indicating the direction to the target. Widely used in aviation, shipping, navigation, weather forecasting, space flight, vehicle collision avoidance systems, and the military.

  - o *Global navigation satellite system* (GNSS) receiver, such as a GPS receiver used with the US Global Positioning System - the most widely used electronic navigation device. An automated digital receiver that receives simultaneous data signals from several satellites in low Earth orbit. Using extremely precise time signals it calculates the distance to the satellites, and from this the receiver's location on Earth. GNSS receivers are sold as portable devices, and are also incorporated in cell phones, vehicles and weapons, even artillery shells.

  - o *VOR* receiver - navigational instrument on an aircraft that uses the VHF signal from VOR navigational beacons between 108 and 117.95 MHz to determine the direction to the beacon very accurately, for air navigation.

  - o *Wild animal tracking* receiver - a receiver with a directional antenna used to track wild animals which have been tagged with a small VHF transmitter, for wildlife management purposes.

- Other

  - o *Telemetry* receiver - this receives data signals to monitor conditions of a process. Telemetry is used to monitor missile and spacecraft in flight, well logging during oil and gas drilling, and unmanned scientific instruments in remote locations.

  - o *Measuring receiver* - a calibrated, laboratory grade radio receiver used to measure the characteristics of radio signals. Often incorporates a spectrum analyzer.

- o   *Radio telescope* - specialized antenna and radio receiver used as a scientific instrument to study weak radio waves from astronomical radio sources in space like stars, nebulas and galaxies in radio astronomy. They are the most sensitive radio receivers that exist, having large parabolic (dish) antennas up to 500 meters in diameter, and extremely sensitive radio circuits. The RF front end of the receiver is often cryogenically cooled by liquid nitrogen to reduce radio noise.

How Receivers Work

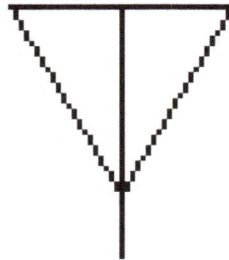

Symbol for an antenna

A radio receiver is connected to an antenna which converts some of the energy from the incoming radio wave into a tiny radio frequency AC voltage which is applied to the receiver's input. An antenna typically consists of an arrangement of metal conductors. The oscillating electric and magnetic fields of the radio wave push the electrons in the antenna back and forth, creating an oscillating voltage.

The antenna may be enclosed inside the receiver's case, as with the ferrite loop antennas of AM radios and the flat inverted F antenna of cell phones; attached to the outside of the receiver, as with whip antennas used on FM radios, or mounted separately and connected to the receiver by a cable, as with rooftop television antennas and satellite dishes.

Filtering, Amplification, and Demodulation

Practical radio receivers perform three basic functions on the signal from the antenna: filtering, amplification, and demodulation:

Symbol for a bandpass filter used in block diagrams of radio receivers

- *Bandpass filtering:* Radio waves from many transmitters pass through the air simultaneously without interfering with each other. These can be separated in the receiver because they have different frequencies. To separate out the desired radio signal, the bandpass filter allows the frequency of the radio transmission to pass though, and blocks signals at all other frequencies.

The bandpass filter consists of one or more resonant circuits (tuned circuits). When the incoming radio signal is at the resonant frequency, the radio signal from the desired station is passed on to the following stages of the receiver. At all other frequencies the tuned circuit has low impedance, so signals at these frequencies are not passed on.

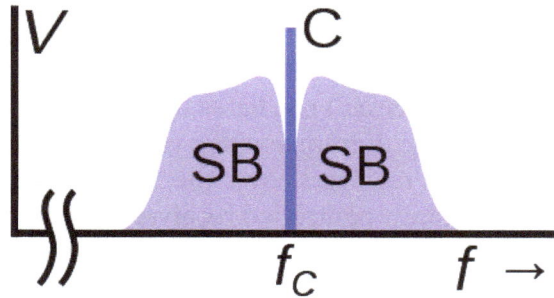

The frequency spectrum of a typical radio signal from an AM or FM radio transmitter. It consists of a strong component (C) at the carrier wave frequency f_c, with the modulation contained in narrow frequency bands called sidebands (SB) just above and below the carrier.

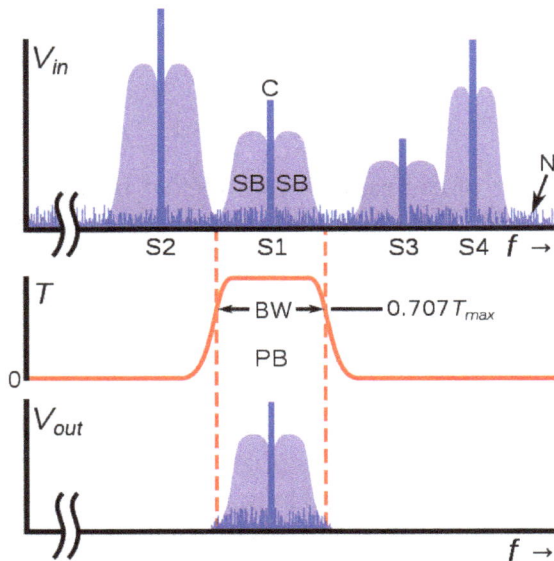

How the bandpass filter selects a single radio signal $S1$ from all the radio signals received by the antenna. From top, the graphs show the voltage from the antenna applied to the filter V_{in}, the transfer function of the filter T, and the voltage at the output of the filter V_{out} as a function of frequency f. The transfer function T is the amount of signal that gets through the filter at each frequency: $V_{out}(f) = T(f)V_{in}(f)$

- *Bandwidth and selectivity:* The information (modulation) in a radio transmission is contained in two narrow bands of frequencies called sidebands *(SB)* on either side of the carrier frequency *(C)*, so the filter has to pass a band of frequencies, not just a single frequency. The band of frequencies received by the receiver is called its *passband (PB)*, and the width of the passband in kilohertz is called the bandwidth *(BW)*. The bandwidth of the filter must be wide enough to allow the sidebands through without distortion, but narrow enough to block any interfering transmissions on adjacent frequencies (such as *S2* in the diagram). The ability of the receiver to reject unwanted radio stations near in frequency to the desired station is an important parameter called *selectivity* determined by the filter. In modern receivers quartz crystal, ceramic resonator, or surface acoustic wave (SAW) filters are often used in place of tuned circuits, which have sharper selectivity.

- *Tuning: Tuning* is adjusting the frequency of the receiver's passband to the frequency of the desired radio transmitter. Tuning a radio has similarities to tuning a musical instrument to resonate with another. Turning the tuning knob changes the resonant frequency of the tuned circuit. When the resonant frequency is equal to the radio transmitter's frequency the tuned circuit oscillates in sympathy, passing the signal on to the rest of the receiver. The range of frequencies a receiver can be tuned to is called its tuning range. For example, FM receivers cover the FM band of frequencies, from 88 MHz to 108 MHz in the US. Digitally controlled receivers use such methods as [phase locked loop]]s to select a desired frequency, and may use a keypad and digital display to show the frequency or a channel number.

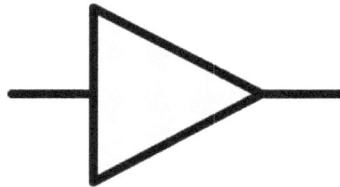

Symbol for an amplifier

- Amplification: The power of the radio waves picked up by a receiving antenna decreases with the square of its distance from the transmitting antenna. Even with the powerful transmitters used in radio broadcasting stations, if the receiver is more than a few miles from the transmitter the power intercepted by the receiver's antenna is very small, perhaps as low as picowatts. To increase the power of the recovered signal, an amplifier turns electric power into a replica of the original signal.

Receivers usually have several stages of amplification: the radio signal from the bandpass filter is amplified to make it powerful enough to drive the demodulator, then the audio signal from the demodulator is amplified to make it powerful enough to operate the speaker. The degree of amplification of a radio receiver is measured by a parameter called its *sensitivity*, which is the minimum signal strength of a station at the antenna, measured in microvolts, necessary to receive the signal clearly, with a certain signal-to-noise ratio. Since it is easy to amplify a signal to any desired degree, the limit to the sensitivity of many modern receivers is not the degree of amplification but random electronic noise present in the circuit, which can drown out a weak radio signal.

Symbol for a demodulator

- *Demodulation:* After the radio signal is filtered and amplified, the receiver must extract the information-bearing modulation signal from the modulated radio frequency carrier wave. This is done by a circuit called a demodulator (detector). Each type of modulation uses a matching type of detector for optimum results; FM, for example, works poorly with an AM detector. Many other types of modulation are also used for specialized purposes. These different types of modulation require different demodulation circuits.

The modulation signal output by the demodulator is usually amplified to increase its strength, then the information is converted back to a human-usable form by some type of transducer. An audio signal, representing sound, as in a broadcast radio, is converted to sound waves by an earphone or loudspeaker. A video signal, representing moving images, as in a television receiver, is converted to light by a display. Digital data, as in a wireless modem, is applied as input to a computer or microprocessor, which interacts with human users.

AM Demodulation

The easiest type of demodulation to understand is AM demodulation, used in AM radios to recover the audio modulation signal, which represents sound and is converted to sound waves by the radio's speaker. It is accomplished by a circuit called an envelope detector, consisting of a diode *(D)* with a bypass capacitor *(C)* across its output.

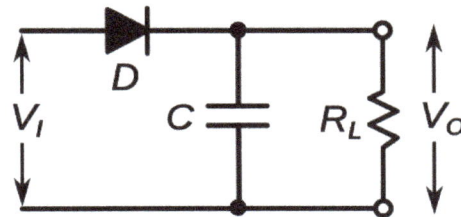

Envelope detector circuit

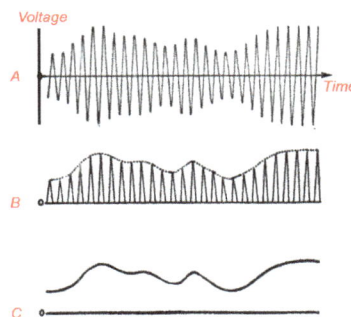

How an envelope detector works

The amplitude modulated radio signal from the tuned circuit is shown at *(A)*. The rapid oscillations are the radio frequency carrier wave. The audio signal (the sound) is contained in the slow variations (modulation) of the amplitude (size) of the waves. If it was applied directly to the speaker, this signal cannot be converted to sound, because the audio excursions are the same on both sides of the axis, averaging out to zero, which would result in no net motion of the speaker's diaphragm. *(B)* When this signal is applied as input V_I to the detector, the diode *(D)* conducts current in one direction but not in the opposite direction, thus allowing through pulses of current on only one side of the signal. In other words, it rectifies the AC current to a pulsing DC current. The resulting voltage V_O applied to the load R_L no longer averages zero; its peak value is proportional to the audio signal. *(C)* The bypass capacitor *(C)* is charged up by the current pulses from the diode, and its voltage follows the peaks of the pulses, the envelope of the audio wave. It performs a smoothing (low pass filtering) function, removing the radio frequency carrier pulses, leaving the low frequency audio signal to pass through the load R_L. The audio signal is amplified and applied to earphones or a speaker.

Tuned Radio Frequency (TRF) Receiver

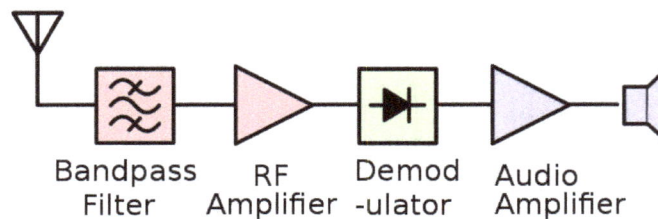

Bandpass   RF   Demod   Audio
Filter   Amplifier -ulator Amplifier

Block diagram of TRF receiver

In the simplest type of radio receiver, called a tuned radio frequency (TRF) receiver, the three functions above are performed consecutively: (1) the mix of radio signals from the antenna is filtered to extract the signal of the desired transmitter; (2) this oscillating voltage is sent through a radio frequency (RF) amplifier to increase its strength to a level sufficient to drive the demodulator; (3) the demodulator recovers the modulation signal (which in broadcast receivers is an audio signal, a voltage oscillating at an audio frequency rate representing the sound waves) from the modulated radio carrier wave; (4) the modulation signal is amplified further in an audio amplifier, then is applied to a loudspeaker or earphone to convert it to sound waves.

Although the TRF receiver is used in a few applications, it has practical disadvantages which make it inferior to the superheterodyne receiver below, which is used in most applications. The drawbacks stem from the fact that in the TRF the filtering, amplification, and demodulation are done at the high frequency of the incoming radio signal. The bandwidth of a filter increases with its center frequency, so as the TRF receiver is tuned to different frequencies its bandwidth varies. Most important, the increasing congestion of the radio spectrum requires that radio channels be spaced very close together in frequency. It is extremely difficult to build filters operating at radio frequencies that have a narrow enough bandwidth to separate closely spaced radio stations. TRF receivers typically must have many cascaded tuning stages to achieve adequate selectivity. The Advantages section below describes how the superheterodyne receiver overcomes these problems.

The Superheterodyne Design

The superheterodyne receiver, invented in 1918 by Edwin Armstrong is the design used in almost all modern receivers except a few specialized applications.

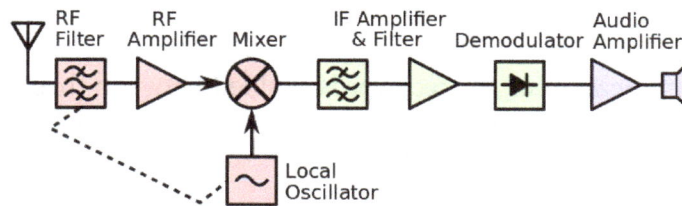

Block diagram of a superheterodyne receiver.

In the superheterodyne, the radio frequency signal from the antenna is shifted down to a lower "intermediate frequency" (IF), before it is processed. The incoming radio frequency signal from the antenna is mixed with an unmodulated signal generated by a *local oscillator* (LO) in the receiver. The mixing is done in a nonlinear circuit called the "*mixer*". The result at the output of the mixer is a heterodyne or beat frequency at the difference between these two frequencies. The process is similar to the way two musical notes at different frequencies played together produce a beat note. This lower frequency is called the *intermediate frequency* (IF). The IF signal also has all the information that was present in the original RF signal. The IF signal passes through filter and amplifier stages, then is demodulated in a detector, recoveirng the original modulation.

The receiver is easy to tune; to receive a different frequency it is only necessary to change the local oscillator frequency. The stages of the receiver after the mixer operates at the fixed intermediate frequency (IF) so the IF bandpass filter does not have to be adjusted to different frequencies. The fixed frequency allows modern receivers to use sophisticated quartz crystal, ceramic resonator, or surface acoustic wave (SAW) IF filters that have very high Q factors, to improve selectivity.

The RF filter on the front end of the receiver is needed to prevent interference from any radio signals at the image frequency. Without an input filter the receiver can receive incoming RF signals at two different frequencies,. The receiver can be designed to receive on either of these two frequencies; if the receiver is designed to receive on one, any other radio station or radio noise on the other frequency may pass through and interfere with the desired signal. A single tunable RF filter stage rejects the image frequency; since these are relatively far from the desired frequency, a simple filter provides adequate rejection. Rejection of interfering signals much closer in frequency to the desired signal is handled by the multiple sharply-tuned stages of the intermediate frequency amplifiers, which do no need to change their tuning. This filter does not need great selectivity, but as the receiver is tuned to different frequencies it must "track" in tandem with the local oscillator. The RF filter also serves to limit the bandwidth applied to the RF amplifier, preventing it from being overloaded by strong out-of-band signals.

To achieve both good image rejection and selectivity, many modern superhet receivers use two intermediate frequencies; this is called a *dual-conversion* or *double-conversion* superheterodyne. The incoming RF signal is first mixed with one local oscillator signal in the first mixer to convert it to a high IF frequency, to allow efficient filtering out of the image frequency, then this first IF is mixed with a second local oscillator signal in a second mixer to convert it to a low IF frequency for good bandpass filtering. Some receivers even use triple-conversion.

At the cost of the extra stages, the superheterodyne receiver provides the advantage of greater selectivity than can be achieved with a TRF design. Where very high frequencies are in use, only the initial stage of the receiver needs to operate at the highest frequencies; the remaining stages can provide much of the receiver gain at lower frequencies which may be easier to manage. Tuning is simplified compared to a multi-stage TRF design, and only two stages need to track over the tuning range. The total amplification of the receiver is divided between three amplifiers at different frequencies; the RF, IF, and audio amplifier. This reduces problems with feedback and parasitic oscillations that are encountered in receivers where most of the amplifier stages operate at the same frequency, as in the TRF receiver.

The most important advantage is that better selectivity can be achieved by doing the filtering at the lower intermediate frequency. One of the most important parameters of a receiver is its bandwidth, the band of frequencies it accepts. In order to reject nearby interfering stations or noise, a narrow bandwidth is required. In all known filtering techniques, the bandwidth of the filter increases in proportion with the frequency, so by performing the filtering at the lower , rather than the frequency of the original radio signal , a narrower bandwidth can be achieved. Modern FM and television broadcasting, cellphones and other communications services, with their narrow channel widths, would be impossible without the superheterodyne.

Automatic Gain Control (AGC)

The signal strength (amplitude) of the radio signal from a receiver's antenna varies drastically, by orders of magnitude, depending on how far away the radio transmitter is, how powerful it is, and propagation conditions along the path of the radio waves. The strength of the signal received from a given transmitter varies with time due to changing propagation conditions of the path through which the radio wave passes, such as multipath interference; this is called *fading*. In an AM receiver the amplitude of the audio signal from the detector, and the sound volume, is proportional to the amplitude of the radio signal, so fading causes variations in the volume. In addition as the receiver is tuned between strong and weak stations, the volume of the sound from the speaker would vary drastically. Without an automatic system to handle it, in an AM receiver constant adjustment of the volume control would be required.

With other types of modulation like FM or FSK the amplitude of the modulation does not vary with the radio signal strength, but in all types the demodulator requires a certain range of signal amplitude to operate properly. Insufficient signal amplitude will cause an increase of noise in the demodulator, while excessive signal amplitude will cause amplifier stages to overload (saturate), causing distortion (clipping) of the signal.

Therefore, almost all modern receivers include a feedback control system which monitors the *average* level of the radio signal at the detector, and adjusts the gain of the amplifiers to give the optimum signal level for demodulation. This is called automatic gain control (AGC). AGC can be compared to the dark adaptation mechanism in the human eye; on entering a dark room the gain of the eye is increased by the iris opening. In its simplest form an AGC system consists of a rectifier which converts the RF signal to a varying DC level, a lowpass filter to smooth the variations and produce an average level. This is applied as a control signal to an earlier amplifier stage, to control its gain. In a superheterodyne receiver AGC is usually applied to the IF amplifier, and there may be a second AGC loop to control the gain of the RF amplifier to prevent it from overloading, too.

In certain receiver designs such as modern digital receivers, a related problem is DC offset of the signal. This is corrected by a similar feedback system.

History

Radio waves were first identified in German physicist Heinrich Hertz's 1887 series of experiments to prove James Clerk Maxwell's electromagnetic theory. Hertz used spark-excited dipole antennas to generate the waves and micrometer spark gaps attached to dipole and loop antennas to detect them. These primitive devices are more accurately described as radio wave sensors, not "receivers", as they could only detect radio waves within about 100 feet of the transmitter, and were not used for communication but instead as laboratory instruments in scientific experiments.

Spark Era

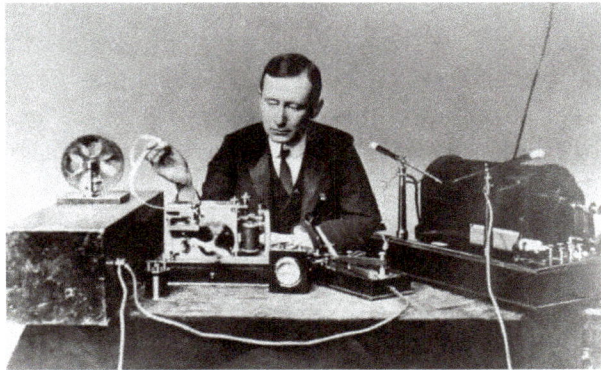

Guglielmo Marconi who built the first radio receivers, with his early spark transmitter *(right)* and coherer receiver *(left)* from the 1890s. The receiver records the Morse code on paper tape

In the early years of radio, from 1887 to 1917, spark gap transmitters generated radio waves by discharging a capacitance through an electric spark. Each spark produced a transient pulse of radio waves which decreased rapidly to zero. These damped waves could not be easily modulated to carry sound, as in modern AM and FM transmission. Spark transmitters could not transmit sound, and instead transmitted information by radiotelegraphy. The transmitter was switched on and off rapidly by the operator using a telegraph key, creating different length pulses of damped radio waves ("dots" and "dashes") to spell out text messages in Morse code.

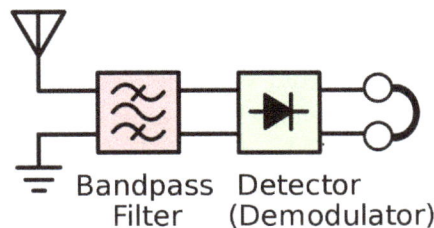

Generic block diagram of an unamplified radio receiver from the wireless telegraphy era

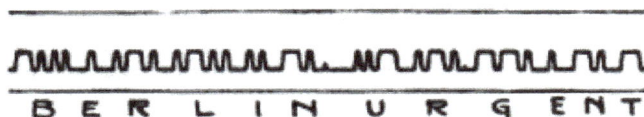

Example of transatlantic radiotelegraph message recorded on paper tape by a siphon recorder at RCA's New York receiving center in 1920. The translation of the Morse code is given below the tape.

The first radio receivers did not have to extract an audio signal but just detected the presence of the radio signal, and produced a sound during the "dots" and "dashes". The device which did this was called a "*detector*". Since there were no amplifying devices, the sensitivity of the receiver mostly depended on the detector. Many different detector devices were tried. Radio receivers during the spark era consisted of these parts:

- An *antenna*, to intercept the radio waves and convert them to tiny radio frequency electric currents.

- A *tuned circuit*, consisting of a capacitor connected to a coil of wire.

, which acted as a bandpass filter to select the desired signal out of all the signals picked up by the antenna. Either the capacitor or coil was adjustable to tune the receiver to the frequency of different transmitters. The earliest receivers, before 1897, did not have tuned circuits, they responded to all radio signals picked up by their antennas, so they had little frequency-discriminating ability and received any transmitter in their vicinity. Most receivers used a pair of tuned circuits with their coils magnetically coupled, called a resonant transformer (oscillation transformer) or "loose coupler".

- A *detector*, which produced a pulse of DC current for each damped wave received.

- An indicating device such as an *earphone*, which converted the pulses of current into sound waves. The first receivers used an electric bell instead. Later receivers in commercial wireless systems used a Morse siphon recorder, which consisted of an ink pen mounted on a needle swung by an electromagnet (a galvanometer) which drew a line on a moving paper tape. Each string of damped waves constituting a Morse "dot" or "dash" caused the needle to swing over, creating a displacement of the line, which could be read off the tape. With such an automated receiver a radio operator didn't have to continuously monitor the receiver.

The signal from the spark gap transmitter consisted of damped waves repeated at an audio frequency rate, from 120 to perhaps 4000 per second, so in the earphone the signal sounded like a musical tone or buzz, and the Morse code "dots" and "dashes" sounded like beeps.

The first person to use radio waves for *communication* was Guglielmo Marconi. Marconi invented little himself, but he was first to believe that radio could be a practical communication medium, and singlehandedly developed the first wireless telegraphy systems, transmitters and receivers, beginning in 1894-5, mainly by improving technology invented by others. Oliver Lodge and Alexander Popov were also experimenting with similar radio wave receiving apparatus at the same time in 1894-5, but they are not known to have transmitted Morse code during this period, just strings of random pulses. Therefore, Marconi is usually given credit for building the first radio receivers.

Coherer Receiver

The first radio receivers invented by Marconi, Oliver Lodge and Alexander Popov in 1894-5 used a primitive radio wave detector called a coherer, invented in 1890 by Edouard Branly and improved by Lodge and Marconi. The coherer was a glass tube with metal electrodes at each end, with loose metal powder between the electrodes. It initially had a high resistance. When a radio frequency voltage was applied to the electrodes, its resistance dropped and it conducted electricity. In the receiver the coherer was connected directly between the antenna and ground. In addition to the antenna, the coherer was connected in a DC circuit with a battery and relay. When the incoming radio wave reduced the resistance of the coherer, the current from the battery flowed through it, turning on the relay to ring a bell or make a mark on a paper tape in a siphon recorder. In order to restore the coherer to its previous nonconducting state to receive the next pulse of radio waves, it had to be tapped mechanically to disturb the metal particles. This was done by a "decoherer", a clapper which struck the tube, operated by an electromagnet powered by the relay.

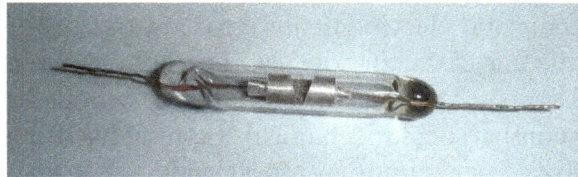

Coherer from 1904 as developed by Marconi.

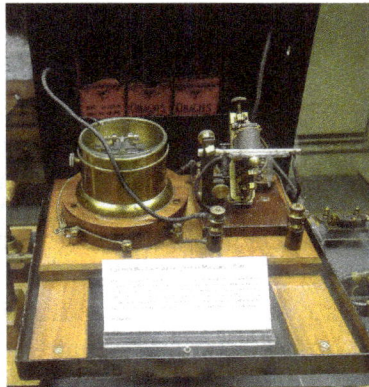

One of Marconi's first coherer receivers, used in his "black box" demonstration at Toynbee Hall, London, 1896. The coherer is at right, with the "tapper" just behind it, The relay is at left, batteries are in background

The coherer is an obscure antique device, and even today there is some uncertainty about the exact physical mechanism by which the various types worked. However it can be seen that it was essentially a bistable device, a radio-wave-operated switch, and so it did not have the ability to rectify the radio wave to demodulate the later amplitude modulated (AM) radio transmissions that carried sound.

A typical commercial radiotelegraphy receiver from the first decade of the 20th century. The coherer (right) detects the pulses of radio waves, and the "dots" and "dashes" of Morse code were recorded in ink on paper tape by a siphon recorder (left) and transcribed later.

In a long series of experiments Marconi found that by using an elevated wire monopole antenna instead of Hertz's dipole antennas he could transmit longer distances, beyond the curve of the Earth, demonstrating that radio was not just a laboratory curiosity but a commercially viable communication method. This culminated in his historic transatlantic wireless transmission on December 12, 1901 from Poldhu, Cornwall to St. John's, Newfoundland, a distance of 3500 km (2200 miles), which was received by a coherer. However the usual range of coherer receivers even with the powerful transmitters of this era was limited to a few hundred miles.

The coherer remained the dominant detector used in early radio receivers for about 10 years, until replaced by the crystal detector and electrolytic detector around 1907. In spite of much development work, it was a very crude unsatisfactory device. It was not very sensitive, and also responded to impulsive radio noise (RFI), such as nearby lights being switched on or off, as well as to the intended signal. Due to the cumbersome mechanical "tapping back" mechanism it was limited to a data rate of about 12-15 words per minute of Morse code, while a spark-gap transmitter could transmit Morse at up to 100 WPM with a paper tape machine.

Other Early Detectors

The coherer's poor performance motivated a great deal of research to find better radio wave detectors, and many were invented. Some strange devices were tried; researchers experimented with using frog legs and even a human brain from a cadaver as detectors.

By the first years of the 20th century, experiments in using amplitude modulation (AM) to transmit sound by radio (radiotelephony) were being made. So a second goal of detector research was to find detectors that could demodulate an AM signal, extracting the audio (sound) signal from the radio carrier wave. It was found by trial and error that this could be done by a detector that exhibited "asymmetrical conduction"; a device that conducted current in one direction but not in the other. This rectified the alternating current radio signal, removing one side of the carrier cycles, leaving a pulsing DC current whose amplitude varied with the audio modulation signal. When applied to an earphone this would reproduce the transmitted sound.

Below are the detectors that saw wide use before vacuum tubes took over around 1920. All except the magnetic detector could rectify and therefore receive AM signals:

Magnetic Detector

- Magnetic detector - Developed by Guglielmo Marconi in 1902 from a method invented by Ernest Rutherford and used by the Marconi Co. until it adopted the Audion vacuum tube around 1912, this was a mechanical device consisting of an endless band of iron wires which passed between two pulleys turned by a windup mechanism. The iron wires passed through a coil of fine wire attached to the antenna, in a magnetic field created by two magnets. The hysteresis of the iron induced a pulse of current in a sensor coil each time a radio signal passed through the exciting coil. The magnetic detector was used on shipboard receivers due to its insensitivity to vibration. One was part of the wireless station of the RMS *Titanic* which was used to summon help during its famous 15 April 1912 sinking.

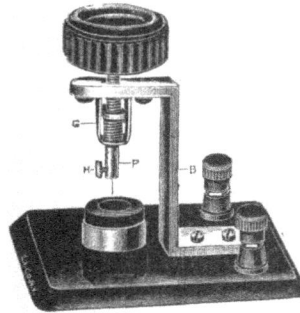

Electrolytic Detector

- Electrolytic detector ("liquid barretter") - Invented in 1903 by Reginald Fessenden, this consisted of a thin silver-plated platinum wire enclosed in a glass rod, with the tip making contact with the surface of a cup of nitric acid. The electrolytic action caused current to be conducted in only one direction. The detector was used until about 1910. Electrolytic detectors that Fessenden had installed on US Navy ships received the first AM radio broadcast on Christmas Eve, 1906, an evening of Christmas music transmitted by Fessenden using his new alternator transmitter.

Early Fleming valve.

Marconi valve receiver for use on ships had two Fleming valves *(top)* in case one burned out. It was used on the RMS *Titanic*.

- Thermionic diode (Fleming valve) - The first vacuum tube, invented in 1904 by John Ambrose Fleming, consisted of an evacuated glass bulb containing two electrodes: a cathode consisting of a hot wire filament similar to that in an incandescent light bulb, and a metal plate anode. Fleming, a consultant to Marconi, invented the valve as a more sensitive detector for transatlantic wireless reception. The filament was heated by a separate current through it and emitted electrons into the tube by thermionic emission, an effect which had been discovered by Thomas Edison. The radio signal was applied between the cathode and anode. When the anode was positive, a current of electrons flowed from the cathode to the anode, but when the anode was negative the electrons were repelled and no current flowed. The Fleming valve was used to a limited extent but was not popular because it was expensive, had limited filament life, and was not as sensitive as electrolytic or crystal detectors.

A galena cat's whisker detector from a 1920s crystal radio

- Crystal detector (cat's whisker detector) - invented around 1904-1906 by Henry H. C. Dunwoody and Greenleaf Whittier Pickard, based on Karl Ferdinand Braun's 1874 discovery of "asymmetrical conduction" in crystals, these were the most successful and widely used detectors before the vacuum tube era and gave their name to the *crystal radio* receiver *(below)*. One of the first semiconductor electronic devices, a crystal detector consisted of a pea-sized pebble of a crystalline semiconductor mineral such as galena (lead sulfide) whose surface was touched by a fine springy metal wire mounted on an adjustable arm. This functioned as a primitive diode which conducted electric current in only one direction. In addition to their use in crystal radios, carborundum crystal detectors were also used in some early vacuum tube radios because they were more sensitive than the vacuum tube grid-leak detector.

During the vacuum tube era, the term "detector" changed from meaning a radio wave detector to mean a demodulator, a device that could extract the audio modulation signal from a radio signal. That is its meaning today.

Tuning

"Tuning" adjusting the frequency of the receiver to the frequency of the desired radio transmission. The first receivers had no tuned circuit, the detector was connected directly between the antenna and ground. Due to the lack of any frequency selective components besides the antenna, the bandwidth of the receiver was equal to the broad bandwidth of the antenna. This was acceptable and even necessary because the first Hertzian spark transmitters also lacked a tuned circuit. Due to the impulsive nature of the spark, the energy of the radio waves was spread over a very wide band of frequencies. To receive enough energy from this wideband signal the receiver had to have a wide bandwidth also.

When more than one spark transmitter was radiating in a given area, their frequencies overlapped, so their signals interfered with each other, resulting in garbled reception. Some method was needed to allow the receiver to select which transmitter's signal to receive. Multiple wavelengths produced by a poorly tuned transmitter caused the signal to "dampen", or die down, greatly reducing the power and range of transmission. In 1892, William Crookes gave a lecture on radio in which he suggested using resonance to reduce the bandwidth of transmitters and receivers. Different transmitters could then be "tuned" to transmit on different frequencies so they didn't interfere. The receiver would also have a resonant circuit (tuned circuit), and could receive a particular transmission by "tuning" its resonant circuit to the same frequency as the transmitter, analogously to tuning a musical instrument to resonance with another. This is the system used in all modern radio.

Tuning was used in Hertz's original experiments and practical application of tuning showed up in the early to mid 1890s in wireless systems not specifically designed for radio communication. Nikola Tesla's March 1893 lecture demonstrating the wireless transmission of power for lighting (mainly by what he thought was ground conduction) included elements of tuning. The wireless lighting system consisted of a spark-excited grounded resonant transformer with a wire antenna which transmitted power across the room to another resonant transformer tuned to the frequency of the transmitter, which lighted a Geissler tube. Use of tuning in free space "Hertzian waves" (radio) was explained and demonstrated in Oliver Lodge's 1894 lectures on Hertz's work. At the time Lodge was demonstrating the physics and optical qualities of radio waves instead of attempting to build a communication system but he would go on to develop methods (patented in 1897) of tuning radio (what he called "syntony"), including using variable inductance to tune antennas.

By 1897 the advantages of tuned systems had become clear, and Marconi and the other wireless researchers had incorporated tuned circuits, consisting of capacitors and inductors connected together, into their transmitters and receivers. The tuned circuit acted like an electrical analog of a tuning fork. It had a high impedance at its resonant frequency, but a low impedance at all other frequencies. Connected between the antenna and the detector it served as a bandpass filter, passing the signal of the desired station to the detector, but routing all other signals to ground. The frequency of the station received f was determined by the capacitance C and inductance L in the tuned circuit:

$$f = \frac{1}{2\pi\sqrt{LC}}$$

Inductive Coupling

In order to reject radio noise and interference from other transmitters near in frequency to the desired station, the bandpass filter (tuned circuit) in the receiver has to have a narrow bandwidth, allowing only a narrow band of frequencies through. The form of bandpass filter that was used in the first receivers, which has continued to be used in receivers until recently, was the double-tuned inductively-coupled circuit, or resonant transformer (oscillation transformer or RF transformer). The antenna and ground were connected to a coil of wire, which was magnetically coupled to a second coil with a capacitor across it, which was connected to the detector. The RF alternating current from the antenna through the primary coil created a magnetic field which induced a current in the secondary coil which fed the detector. Both primary and secondary were tuned circuits; the primary coil resonated with the capacitance of the antenna, while the secondary coil resonated with the capacitor across it. Both were adjusted to the same resonant frequency.

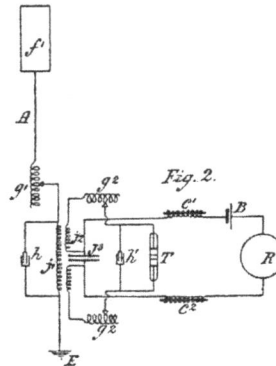

Marconi's inductively coupled coherer receiver from his controversial April 1900 "four circuit" patent no. 7,777.

Braun receiving transformer from 1904

This circuit had two advantages. One was that by using the correct turns ratio, the impedance of the antenna could be matched to the impedance of the receiver, to transfer maximum RF power to the receiver. Impedance matching was important to achieve maximum receiving range in the unamplified receivers of this era. The coils usually had taps which could be selected by a multiposition switch. The second advantage was that due to "loose coupling" it had a much narrower bandwidth than a simple tuned circuit, and the bandwidth could be adjusted. Unlike in an ordinary transformer, the two coils were "loosely coupled"; separated physically so not all the magnetic field from the primary passed through the secondary, reducing the mutual inductance. This gave the coupled tuned circuits much "sharper" tuning, a narrower bandwidth than a single tuned circuit. In the "Navy type" loose coupler, widely used with crystal receivers, the smaller secondary coil was mounted on a rack which could be slid in or out of the primary coil, to vary the mutual inductance between the coils. When the operator encountered an interfering signal at a nearby frequency, the secondary could be slid further out of the primary, reducing the coupling, which narrowed the bandwidth, rejecting the interfering signal. A disadvantage was that all three adjustments in the loose coupler - primary tuning, secondary tuning, and coupling - were interactive; changing one changed the others. So tuning in a new station was a process of successive adjustments.

Crystal receiver from 1914 with "loose coupler" tuning transformer. The secondary coil *(1)* can be slid in or out of the primary *(in box)* to adjust the coupling. Other components: *(2)* primary tuning capacitor, *(3)* secondary tuning capacitor, *(4)* loading coil, *(5)* crystal detector, *(8)* headphones

Selectivity became more important as spark transmitters were replaced by continuous wave transmitters which transmitted on a narrow band of frequencies, and broadcasting led to a proliferation of closely spaced radio stations crowding the radio spectrum. Resonant transformers continued to be used as the bandpass filter in vacuum tube radios, and new forms such as the *variometer* were invented. Another advantage of the double-tuned transformer for AM reception was that when properly adjusted it had a "flat top" frequency response curve as opposed to the "peaked" response of a single tuned circuit. This allowed it to pass the sidebands of AM modulation on either side of the carrier with little distortion, unlike a single tuned circuit which attenuated the higher audio frequencies. Until recently the bandpass filters in the superheterodyne circuit used in all modern receivers were made with resonant transformers, called IF transformers.

Patent Disputes

Marconi's initial radio system had relatively poor tuning limiting its range and adding to interference. To overcome this drawback he developed a four circuit system with tuned coils in "*symphony*" at both the transmitters and receivers. His 1900 British #7,777 (four sevens) patent for tuning filed in April 1900 and granted a year later opened the door to patents disputes since it infringed on the Syntonic patents of Oliver Lodge, first filed in May 1897, as well as patents filed by Ferdinand Braun. Marconi was able to obtain patents in the UK and France but the US version of his tuned four circuit patent, filed in November 1900, was initially rejected based on it being anticipated by Lodge's tuning system, and refiled versions were rejected because of the prior patents by Braun, and Lodge. A further clarification and re-submission was rejected because it infringed on parts of two prior patents Tesla had obtained for his wireless power transmission system. Marconi's lawyers managed to get a resubmitted patent reconsidered by another examiner who initially rejected it due to a pre-existing John Stone Stone tuning patent, but it was finally approved it in June 1904 based on it having a unique system of variable inductance tuning that was different from Stone who tuned by varying the length of the antenna. When Lodge's Syntonic patent was extended in 1911 for another 7 years the Marconi Company agreed to settle that patent dispute, purchasing Lodge's radio company with its patent in 1912, giving them the priority patent they needed. Other patent disputes would crop up over the years including a 1943 US Supreme Court ruling on the Marconi Companies ability to sue the US government over patent infringement during World War I. The

Court rejected the Marconi Companies suit saying they could not sue for patent infringement when their own patents did not seem to have priority over the patents of Lodge, Stone, and Tesla.

Crystal Radio Receiver

Prior to 1920 the crystal receiver was the main type used in wireless telegraphy stations, and sophisticated models were made, like this Marconi Type 106 from 1915.

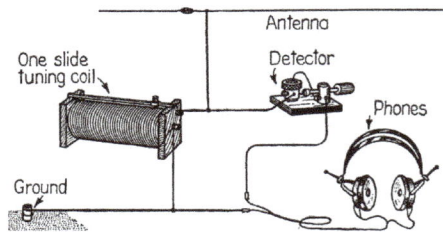

After vacuum tube receivers appeared around 1920, the crystal set became a simple cheap alternative radio used by youth and the poor.

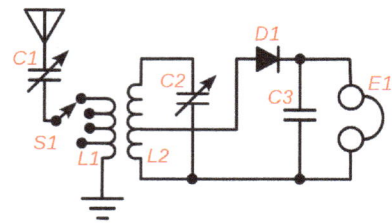

Simple crystal radio. The capacitance of the wire antenna connected to the coil serves as the capacitor in the tuned circuit.

Typical "loose coupler" crystal radio circuit

Although it was invented in 1904 in the wireless telegraphy era, the crystal radio receiver could also rectify AM transmissions and served as a bridge to the broadcast era. In addition to being the main type used in commercial stations during the wireless telegraphy era, it was the first receiver to be used widely by the public. During the first two decades of the 20th century, as radio stations began to transmit in AM voice (radiotelephony) instead of radiotelegraphy, radio listening became a popular hobby, and the crystal was the simplest, cheapest detector. The millions of people who purchased or homemade these inexpensive reliable receivers created the mass listening audience for the first radio broadcasts, which began around 1920. By the late 1920s the crystal receiver was superseded by vacuum tube receivers and became commercially obsolete. However it continued to be used by youth and the poor until World War 2. Today these simple radio receivers are constructed by students as educational science projects.

The crystal radio used a cat's whisker detector, invented by Harrison H. C. Dunwoody and Greenleaf Whittier Pickard in 1904, to extract the audio from the radio frequency signal. It consisted of a mineral crystal, usually galena, which was lightly touched by a fine springy wire (the "cat whisker") on an adjustable arm. The resulting crude semiconductor junction functioned as a Schottky barrier diode, conducting in only one direction. Only particular sites on the crystal surface worked as detector junctions, and the junction could be disrupted by the slightest vibration. So a usable site was found by trial and error before each use; the operator would drag the cat's whisker across the crystal until the radio began functioning. Frederick Seitz, a later semiconductor researcher, wrote:

Such variability, bordering on what seemed the mystical, plagued the early history of crystal detectors and caused many of the vacuum tube experts of a later generation to regard the art of crystal rectification as being close to disreputable.

The crystal radio was unamplified and ran off the power of the radio waves received from the radio station, so it had to be listened to with earphones; it could not drive a loudspeaker. It required a long wire antenna, and its sensitivity depended on how large the antenna was. During the wireless era it was used in commercial and military longwave stations with huge antennas to receive long distance radiotelegraphy traffic, even including transatlantic traffic. However, when used to receive broadcast stations a typical home crystal set had a more limited range of about 25 miles. In sophisticated crystal radios the "loose coupler" inductively coupled tuned circuit was used to increase the Q. However it still had poor selectivity compared to modern receivers.

Heterodyne Receiver and BFO

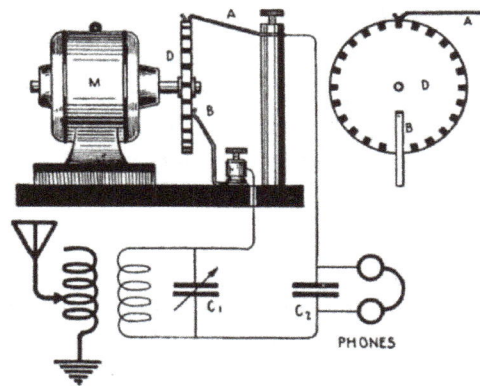

Radio receiver with Poulsen "tikker" consisting of a commutator disk turned by a motor to interrupt the carrier.

Beginning around 1905 continuous wave (CW) transmitters began to replace spark transmitters for radiotelegraphy because they had much greater range. The first continuous wave transmitters were the Poulsen arc invented in 1904 and the Alexanderson alternator developed 1906-1910, which were replaced by vacuum tube transmitters beginning around 1920.

The continuous wave radiotelegraphy signals produced by these transmitters required a different method of reception. The radiotelegraphy signals produced by spark gap transmitters consisted of strings of damped waves repeating at an audio rate, so the "dots" and "dashes" of Morse code were audible as a tone or buzz in the receivers' earphones. However the new continuous wave radiotelegraph signals simply consisted of pulses of unmodulated carrier (sine waves). These were inaudible in the receiver headphones. To receive this new modulation type, the receiver had to produce some kind of tone during the pulses of carrier.

The first crude device that did this was the "ticker" or "tikker", invented in 1908 by Valdemar Poulsen. This was a vibrating interrupter with a capacitor at the tuner output which served as a rudimentary modulator, interrupting the carrier at an audio rate, thus producing a buzz in the earphone when the carrier was present. A similar device was the "tone wheel" invented by Rudolph Goldschmidt, a wheel spun by a motor with contacts spaced around its circumference, which made contact with a stationary brush.

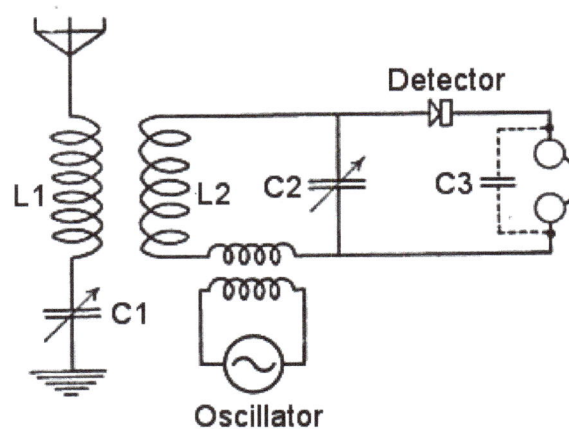

Fessenden's heterodyne radio receiver circuit

In 1901 Reginald Fessenden had invented a better means of accomplishing this. In his *heterodyne receiver* an unmodulated sine wave radio signal at a frequency f_o offset from the incoming radio wave carrier f_c was applied to a rectifying detector such as a crystal detector or electrolytic detector, along with the radio signal from the antenna. In the detector the two signals mixed, creating two new *heterodyne* (beat) frequencies at the sum $f_c + f_o$ and the difference $f_c - f_o$ between these frequencies. By choosing f_o correctly the lower heterodyne $f_c - f_o$ was in the audio frequency range, so it was audible as a tone in the earphone whenever the carrier was present. Thus the "dots" and "dashes" of Morse code were audible as musical "beeps". A major attraction of this method during this pre-amplification period was that the heterodyne receiver actually amplified the signal somewhat, the detector had "mixer gain".

The receiver was ahead of its time, because when it was invented there was no oscillator capable of producing the radio frequency sine wave f_o with the required stability. Fessenden first used his large radio frequency alternator, but this wasn't practical for ordinary receivers. The heterodyne receiver remained a laboratory curiosity until a cheap compact source of continuous waves appeared, the vacuum tube electronic oscillator invented by Edwin Armstrong and Alexander Meissner in 1913. After this it became the standard method of receiving CW radiotelegraphy. The heterodyne oscillator is the ancestor of the *beat frequency oscillator* (BFO) which is used to receive radiotelegraphy in communications receivers today. The heterodyne oscillator had to be retuned each time the receiver was tuned to a new station, but in modern superheterodyne receivers the BFO signal beats with the fixed intermediate frequency, so the beat frequency oscillator can be a fixed frequency.

Armstrong later used Fessenden's heterodyne principle in his superheterodyne receiver *(below)*.

Vacuum Tube Era

The Audion (triode) vacuum tube invented by Lee De Forest in 1906 was the first practical amplifying device and revolutionized radio. Vacuum tube transmitters replaced spark transmitters and made possible four new types of modulation: continuous wave (CW) radiotelegraphy, amplitude modulation (AM) around 1915 which could carry audio (sound), frequency modulation (FM) around 1938 which had much improved audio quality, and single sideband (SSB).

SUPER-HETERODYNE

THE TROPADYNE

THE NEUTRODYNE

5-TUBE COCKADAY

THE REFLEX

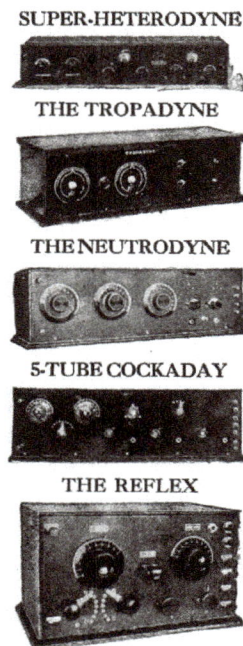

Unlike today, when almost all radios use a variation of the superheterodyne design, during the 1920s vacuum tube radios used a variety of competing circuits.

During the "Golden Age of Radio" (1920 to 1950), families gathered to listen to the home radio in the evening, such as this Zenith console model 12-S-568 from 1938, a 12 tube superheterodyne with pushbutton tuning and 12 inch cone speaker.

The amplifying vacuum tube used energy from a battery or electrical outlet to increase the power of the radio signal, so vacuum tube receivers could be more sensitive and have a greater reception range than the previous unamplified receivers. The increased audio output power also allowed them to drive loudspeakers instead of earphones, permitting more than one person to listen. The first loudspeakers were produced around 1915. These changes caused radio listening to evolve explosively from a solitary hobby to a popular social and family pastime. The development of amplitude modulation (AM) and vacuum tube transmitters during World War I, and the availability of cheap receiving tubes after the war, set the stage for the start of AM broadcasting, which sprang up spontaneously around 1920.

The advent of radio broadcasting increased the market for radio receivers greatly, and transformed them into a consumer product. At the beginning of the 1920s the radio receiver was a forbidding high-tech device, with many cryptic knobs and controls requiring technical skill to operate, housed in an unattractive black metal box, with a tinny-sounding horn loudspeaker. By the 1930s, the broadcast receiver had become a piece of furniture, housed in an attractive wooden case, with standardized controls anyone could use, which occupied a respected place in the home living room. In the early radios the multiple tuned circuits required multiple knobs to be adjusted to tune in a new station. One of the most important ease-of-use innovations was "single knob tuning", achieved by linking the tuning capacitors together mechanically. The dynamic cone loudspeaker invented in 1924 greatly improved audio frequency response over the previous horn speakers, allowing music to be reproduced with good fidelity. Convenience features like large lighted dials, tone controls, pushbutton tuning, tuning indicators and automatic gain control (AGC) were added. The receiver market was divided into the above *broadcast receivers* and *communications receivers*, which were used for two-way radio communications such as shortwave radio.

A vacuum tube receiver required several power supplies at different voltages, which in early radios were supplied by separate batteries. By 1930 adequate rectifier tubes were developed, and the expensive batteries were replaced by a transformer power supply that worked off the house current.

Vacuum tubes were bulky, expensive, had a limited lifetime, consumed a large amount of power and produced a lot of waste heat, so the number of tubes a receiver could economically have was a limiting factor. Therefore, a goal of tube receiver design was to get the most performance out of a limited number of tubes. The major radio receiver designs, listed below, were invented during the vacuum tube era.

A defect in many early vacuum tube receivers was that the amplifying stages could oscillate, act as an oscillator, producing unwanted radio frequency alternating currents. These parasitic oscillations mixed with the carrier of the radio signal in the detector tube, producing audible beat notes (heterodynes); annoying whistles, moans, and howls in the speaker. The oscillations were caused by feedback in the amplifiers; one major feedback path was the capacitance between the plate and grid in early triodes. This was solved by the Neutrodyne circuit, and later the development of the tetrode and pentode around 1930.

Edwin Armstrong is one of the most important figures in radio receiver history, and during this period invented technology which continues to dominate radio communication. He was the first to give a correct explanation of how De Forest's triode tube worked. He invented the feedback oscillator, regenerative receiver, the superregenerative receiver, the superheterodyne receiver, and modern frequency modulation (FM).

The First Vacuum Tube Receivers

The first amplifying vacuum tube, the Audion, a crude triode, was invented in 1906 by Lee De Forest as a more sensitive detector for radio receivers, by adding a third electrode to the thermionic diode detector, the Fleming valve. It was not widely used until its amplifying ability was recognized around 1912. The first tube receivers, invented by De Forest and built by hobbyists until the mid 1920s, used a single Audion which functioned as a grid-leak detector which both rectified and amplified the radio signal. There was uncertainty about the operating principle of the Audion until

Edwin Armstrong explained both its amplifying and demodulating functions in a 1914 paper. The grid-leak detector circuit was also used in regenerative, TRF, and early superheterodyne receivers *(below)* until the 1930s.

De Forest's first commercial Audion receiver, the RJ6 which came out in 1914. The Audion tube was always mounted upside down, with its delicate filament loop hanging down, so it did not sag and touch the other electrodes in the tube.

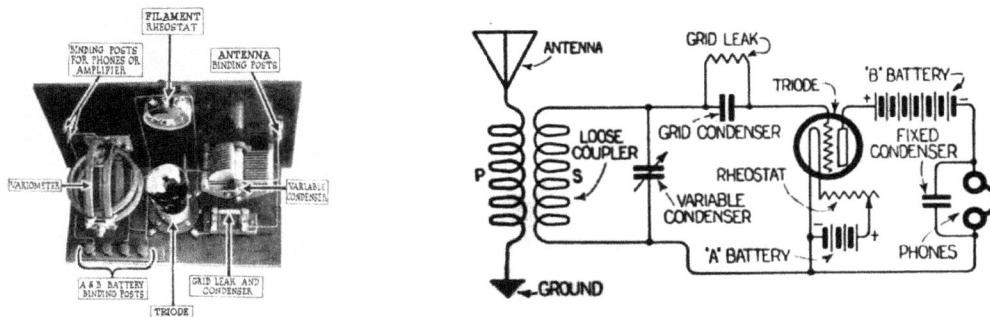

Example of single tube triode grid-leak receiver from 1920, the first type of amplifying radio receiver. In the grid leak circuit, electrons attracted to the grid during the positive half cycles of the radio signal charge the grid capacitor with a negative voltage of a few volts, biasing the grid near its cutoff voltage, so the tube conducts only during the positive half-cycles, rectifying the radio carrier.

To give enough output power to drive a loudspeaker, 2 or 3 additional Audion stages were needed for audio amplification. Many early hobbyists could only afford a single tube receiver, and listened to the radio with earphones, so early tube amplifiers and speakers were sold as add-ons.

In addition to very low gain of about 5 and a short lifetime of about 30 - 100 hours, the primitive Audion had erratic characteristics because it was incompletely evacuated. De Forest believed that ionization of residual air was key to Audion operation. This made it a more sensitive detector but also caused its electrical characteristics to vary during use. As the tube heated up, gas released from the metal elements would change the pressure in the tube, changing the plate current and other characteristics, so it required periodic bias adjustments to keep it at the correct operating point. Each Audion stage usually had a rheostat to adjust the filament current, and often a potentiometer or multiposition switch to control the plate voltage. The filament rheostat was also used as a volume control. The many controls made multitube Audion receivers complicated to operate.

By 1914, Harold Arnold at Western Electric and Irving Langmuir at GE realized that the residual gas was not necessary; the Audion could operate on electron conduction alone. They evacuated tubes to a lower pressure of 10^{-9} atm, producing the first "hard vacuum" triodes. These more stable tubes did not require bias adjustments, so radios had fewer controls and were easier to operate.

During World War I civilian radio use was prohibited, but by 1920 large-scale production of vacuum tube radios began. The "soft" incompletely evacuated tubes were used as detectors through the 1920s then became obsolete.

Regenerative (Autodyne) Receiver

The regenerative receiver, invented by Edwin Armstrong in 1913 when he was a 23-year-old college student, was used very widely until the late 1920s particularly by hobbyists who could only afford a single-tube radio. Today transistor versions of the circuit are still used in a few inexpensive applications like walkie-talkies. In the regenerative receiver the gain (amplification) of a vacuum tube or transistor is increased by using *regeneration* (positive feedback); some of the energy from the tube's output circuit is fed back into the input circuit with a feedback loop. The early vacuum tubes had very low gain (around 5). Regeneration could not only increase the gain of the tube enormously, by a factor of 15,000 or more, it also increased the Q factor of the tuned circuit, decreasing (sharpening) the bandwidth of the receiver by the same factor, improving selectivity greatly. The receiver had a control to adjust the feedback. The tube also acted as a grid-leak detector to rectify the AM signal.

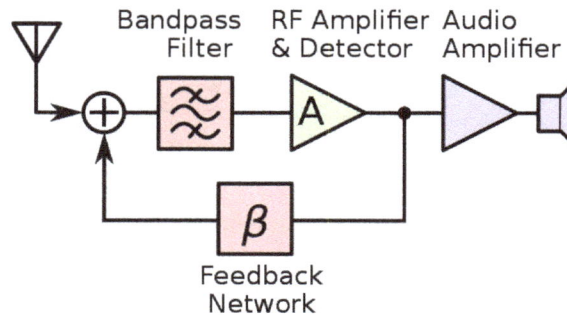

Block diagram of regenerative receiver

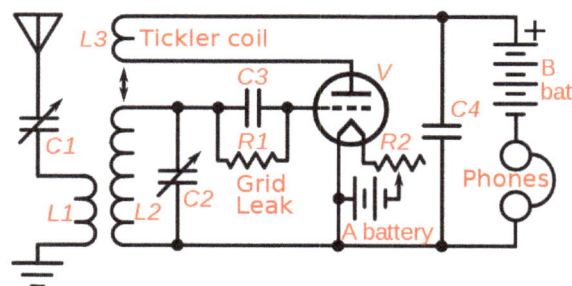

Circuit of single tube Armstrong regenerative receiver

Another advantage of the circuit was that the tube could be made to oscillate, and thus a single tube could serve as both a beat frequency oscillator and a detector, functioning as a heterodyne receiver to make CW radiotelegraphy transmissions audible. This mode was called an autodyne receiver. To receive radiotelegraphy, the feedback was increased until the tube oscillated, then the oscillation frequency was tuned to one side of the transmitted signal. The incoming radio carrier signal and local oscillation signal mixed in the tube and produced an audible heterodyne (beat) tone at the difference between the frequencies.

Homemade Armstrong regenerative receiver, 1922. The "tickler" coil *(L3)* is visible on the front panel, coupled to the input tuning coils.

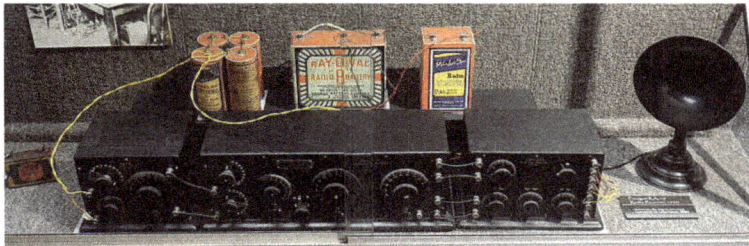

Commercial regenerative receiver from the early 1920s, the Paragon RA-10 *(center)* with separate 10R single tube RF amplifier *(left)* and three tube DA-2 detector and 2-stage audio amplifier unit *(right)*. The 4 cylindrical dry cell "A" batteries *(right rear)* powered the tube filaments, while the 2 rectangular "B" batteries provided plate voltage.

A widely used design was the Armstrong circuit, in which a "tickler" coil in the plate circuit was coupled to the tuning coil in the grid circuit, to provide the feedback. The feedback was controlled by a variable resistor, or alternately by moving the two windings physically closer together to increase loop gain, or apart to reduce it. This was done by an adjustable air core transformer called a variometer (variocoupler). Regenerative detectors were sometimes also used in TRF and superheterodyne receivers.

Homemade one-tube Armstrong regenerative receiver from the 1940s. The tickler coil is a variometer winding mounted on a shaft inside the tuning coil *(upper right)* which can be rotated by a knob on the front panel.

One problem with the regenerative circuit was that when used with large amounts of regeneration the selectivity (Q) of the tuned circuit could be *too* sharp, attenuating the AM sidebands, thus distorting the audio modulation. This was usually the limiting factor on the amount of feedback that could be employed.

A more serious drawback was that it could act as an inadvertent radio transmitter, producing interference (RFI) in nearby receivers. In AM reception, to get the most sensitivity the tube was operated very close to instability and could easily break into oscillation (and in CW reception *did* oscillate), and the resulting radio signal was radiated by its wire antenna. In nearby receivers, the regenerative's signal would beat with the signal of the station being received in the detector, creating annoying heterodynes, (beats), howls and whistles. Early regeneratives which oscillated easily were called "bloopers", and were made illegal in Europe. One preventative measure was to use a stage of RF amplification before the regenerative detector, to isolate it from the antenna. But by the mid 1920s "regens" were no longer sold by the major radio manufacturers.

Superregenerative Receiver

Armstrong presenting his superregenerative receiver, June 28, 1922, Columbia University

This was a receiver invented by Edwin Armstrong in 1922 which used regeneration in a more sophisticated way, to give greater gain. It was used in a few shortwave receivers in the 1930s, and is used today in a few cheap high frequency applications such as walkie-talkies and garage door openers.

In the regenerative receiver the loop gain of the feedback loop was less than one, so the tube (or other amplifying device) did not oscillate but was close to oscillation, giving large gain. In the superregenerative receiver, the loop gain was made equal to one, so the amplifying device actually began to oscillate, but the oscillations were interrupted periodically. This allowed a single tube to produce gains of over 10^6.

TRF Receiver

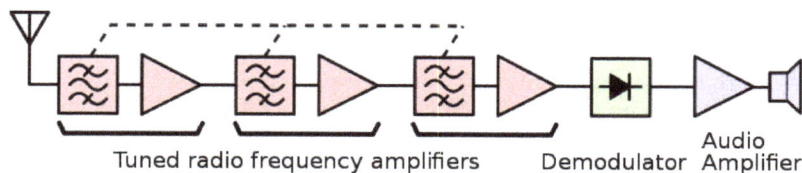

Block diagram of a tuned radio frequency receiver. To achieve enough selectivity to reject stations on adjacent frequencies, multiple cascaded bandpass filter stages had to be used. The dotted line indicates that the bandpass filters must be tuned together.

The tuned radio frequency (TRF) receiver, invented in 1916 by Ernst Alexanderson, improved both sensitivity and selectivity by using several stages of amplification before the detector, each with a tuned circuit, all tuned to the frequency of the station. Its operation is briefly described in the How receivers work section above. It was very popular in quality radios during the 1920s until the superheterodyne replaced it in the 1930s. TRF receivers consist of these parts:

- One or more tuned radio frequency amplifier stages, each consisting of an amplifying tube or transistor and a tuned circuit. In vacuum tube radios these consisted of a tube amplifier followed by an air core interstage coupling transformer with a capacitor across one winding.

- A detector stage, in tube radios usually a triode grid-leak detector.

- Usually one or more audio amplifier stages.

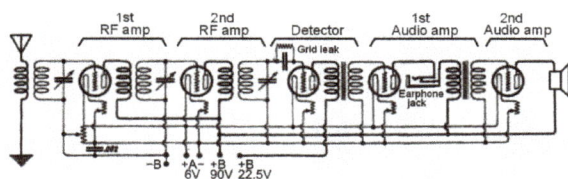

Typical 5 tube TRF circuit from 1924 has 2 stages of RF amplification containing 3 bandpass filters, a grid-leak detector stage, and 2 stages of transformer-coupled audio amplification

Early 6 tube TRF receiver from around 1920. The 3 large knobs adjust the 3 tuned circuits to tune in stations

Atwater-Kent TRF receiver from the 1920s with 2 RF stages *(left)*, detector and two audio amplifier tubes *(right)*. The loudspeaker consists of an earphone coupled to an acoustic horn which amplifies the sound.

Tuning a Neutrodyne TRF receiver with 3 tuned circuits *(large knobs)*, 1924. For each station the index numbers on the dials had to be written down so that the station could be found again.

A major problem of early TRF receivers was that they were complicated to tune, because each resonant circuit had to be adjusted to the frequency of the station before the radio would work. In later TRF receivers the tuning capacitors were linked together mechanically ("ganged") on a common shaft so they could be adjusted with one knob, but in early receivers the frequencies of the tuned circuits could not be made to "track" well enough to allow this, and each tuned circuit had its own tuning knob. Therefore, the knobs had to be turned simultaneously. For this reason most TRF sets had no more than three tuned RF stages.

A second problem was that the multiple radio frequency stages, all tuned to the same frequency, were prone to oscillate, and the parasitic oscillations mixed with the radio station's carrier in the detector, producing audible heterodynes (beat notes), whistles and moans, in the speaker. This was solved by the invention of the Neutrodyne circuit and the development of the tetrode later around 1930, and better shielding between stages.

Today the TRF design is used in a few integrated (IC) receiver chips. From the standpoint of modern receivers the disadvantage of the TRF is that the gain and bandwidth of the tuned RF stages are not constant but vary as the receiver is tuned to different frequencies. Since the bandwidth of a filter with a given Q is proportional to the frequency, as the receiver is tuned to higher frequencies its bandwidth increases.

Neutrodyne Receiver

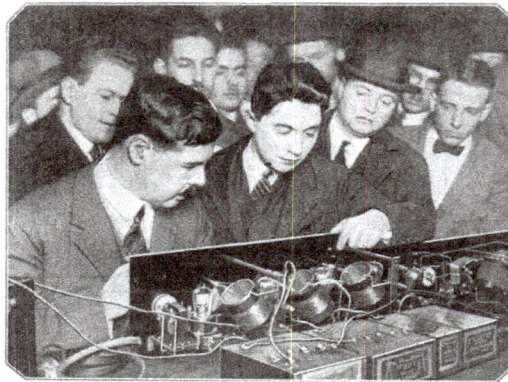

Hazeltine's prototype Neutrodyne receiver, presented at a March 2, 1923 meeting of the Radio Society of America at Columbia University.

The Neutrodyne receiver, invented in 1922 by Louis Hazeltine, was a TRF receiver with a "neutralizing" circuit added to each radio amplification stage to cancel the feedback to prevent the oscillations which caused the annoying whistles in the TRF. In the neutralizing circuit a capacitor fed a feedback current from the plate circuit to the grid circuit which was 180° out of phase with the feedback which caused the oscillation, canceling it. The Neutrodyne was popular until the advent of cheap tetrode tubes around 1930.

Reflex Receiver

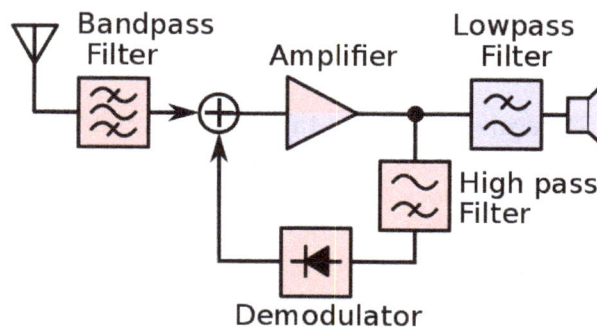

Block diagram of simple single tube reflex receiver

The reflex receiver, invented in 1914 by Wilhelm Schloemilch and Otto von Bronk, and rediscovered and extended to multiple tubes in 1917 by Marius Latour and William H. Priess, was a design used in some inexpensive radios of the 1920s which enjoyed a resurgence in small portable tube radios of the 1930s and again in a few of the first transistor radios in the 1950s. It is another example of an ingenious circuit invented to get the most out of a limited number of active devices. In the reflex receiver the RF signal from the tuned circuit is passed through one or more amplifying tubes or transistors, demodulated in a detector, then the resulting audio signal is passed *again* though the same amplifier stages for audio amplification. The separate radio and audio signals present simultaneously in the amplifier do not interfere with each other since they are at different frequencies, allowing the amplifying tubes to do "double duty". In addition to single tube reflex receivers, some TRF and superheterodyne receivers had several stages "reflexed". Reflex radios were prone to a defect called "play-through" which meant that the volume of audio did not go to zero when the volume control was turned down.

Superheterodyne Receiver

The superheterodyne, invented in 1918 during World War I by Edwin Armstrong when he was in the Signal Corps, is the design used in almost all modern receivers, except a few specialized applications. It is a more complicated design than the other receivers above, and when it was invented required 6 - 9 vacuum tubes, putting it beyond the budget of most consumers, so it was initially used mainly in commercial and military communication stations. However, by the 1930s the "superhet" had replaced all the other receiver types above.

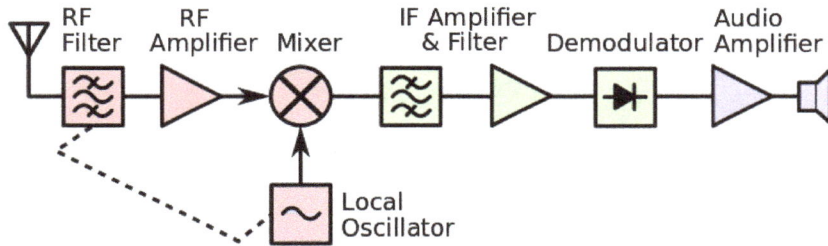

Block diagram of a superheterodyne receiver. The dotted line indicates that the RF filter and local oscillator must be tuned in tandem.

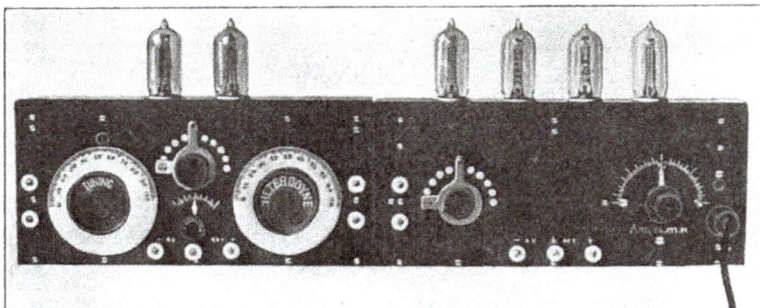

The first superheterodyne receiver built at Armstrong's Signal Corps laboratory in Paris during World War I. It is constructed in two sections, the mixer and local oscillator *(left)* and three IF amplification stages and a detector stage *(right)*. The intermediate frequency was 75 kHz.

In the superheterodyne, the "heterodyne" technique invented by Reginald Fessenden is used to shift the frequency of the radio signal down to a lower "intermediate frequency" (IF), before it is processed.

During the 1940s the vacuum tube superheterodyne receiver was refined into a cheap-to-manufacture form called the "All American Five" because it only required 5 tubes, which was used in almost all broadcast radios until the end of the tube era in the 1970s.

By the 1940s the superheterodyne AM broadcast receiver was refined into a cheap-to-manufacture design called the "All American Five", because it only used five vacuum tubes: usually a converter (mixer/local oscillator), an IF amplifier, a detector/audio amp, audio power amp, and a rectifier. This design was used for virtually all commercial radio receivers until the transistor replaced the vacuum tube in the 1970s.

Semiconductor Era

A modern smartphone has several digital radio transmitters and receivers to connect to different devices; a cellular receiver, a wireless modem, a bluetooth modem, and a GPS receiver

The invention of the transistor in 1947 revolutionized radio technology, making truly portable receivers possible, beginning with transistor radios in the late 1950s. Although portable vacuum tube radios were made, tubes were bulky and inefficient, consuming large amounts of power and requiring several large batteries to produce the filament and plate voltage. Transistors did not require a heated filament, reducing power consumption, and were smaller and much less fragile than vacuum tubes.

Digital Technologies

The development of integrated circuits (ICs) in the 1970s created another revolution, allowing an entire radio receiver to be put on a chip. ICs reversed the economics of radio design used with vacuum tube receivers. Since the marginal cost of adding additional amplifying devices (transistors) to the chip was essentially zero, the size and cost of the receiver was dependent not on how many active components were used, but on the passive components; inductors and capacitors, which could not be integrated easily on the chip.

As a result, the current trend in receivers is to use digital circuitry on the chip to do functions that were formerly done by analog circuits which require passive components. In a digital receiver the IF signal is sampled and digitized, and the bandpass filtering and detection functions are performed by digital signal processing (DSP) on the chip. Another benefit of DSP is that the properties of the receiver; channel frequency, bandwidth, gain, etc. can be dynamically changed by software to react to changes in the environment; these systems are known as software-defined radios or cognitive radio.

Many of the functions performed by analog electronics can be performed by software instead. The benefit is that software is not affected by temperature, physical variables, electronic noise and manufacturing defects.

Digital signal processing permits signal processing techniques that would be cumbersome, costly, or otherwise infeasible with analog methods. A digital signal is essentially a stream or sequence of numbers that relay a message through some sort of medium such as a wire. DSP hardware can tailor the bandwidth of the receiver to current reception conditions and to the type of signal. A typical analog only receiver may have a limited number of fixed bandwidths, or only one, but a DSP receiver may have 40 or more individually selectable filters. DSP is used in cell phone systems to reduce the data rate required to transmit voice.

"PC radios", or radios that are designed to be controlled by a standard PC are controlled by specialized PC software using a serial port connected to the radio. A "PC radio" may not have a front-panel at all, and may be designed exclusively for computer control, which reduces cost.

Some PC radios have the great advantage of being field upgradable by the owner. New versions of the DSP firmware can be downloaded from the manufacturer's web site and uploaded into the flash memory of the radio. The manufacturer can then in effect add new features to the radio over time, such as adding new filters, DSP noise reduction, or simply to correct bugs.

A full-featured radio control program allows for scanning and a host of other functions and, in particular, integration of databases in real-time, like a "TV-Guide" type capability. This is particularly helpful in locating all transmissions on all frequencies of a particular broadcaster, at any given time. Some control software designers have even integrated Google Earth to the shortwave databases, so it is possible to "fly" to a given transmitter site location with a click of a mouse. In many cases the user is able to see the transmitting antennas where the signal is originating from.

Since the Graphical User Interface to the radio has considerable flexibility, new features can be added by the software designer. Features that can be found in advanced control software programs today include a band table, GUI controls corresponding to traditional radio controls, local time clock and a UTC clock, signal strength meter, a database for shortwave listening with lookup capability, scanning capability, or text-to-speech interface.

The next level in integration is "software-defined radio", where all filtering, modulation and signal manipulation is done in software. This may be a PC soundcard or by a dedicated piece of DSP hardware. There will be a RF front-end to s upply an intermediate frequency to the software defined radio. These systems can provide additional capability over "hardware" receivers. For example, they can record large swaths of the radio spectrum to a hard drive for "playback" at a later date. The same SDR that one minute is demodulating a simple AM broadcast may also be able to decode an HDTV broadcast in the next. An open-source project called GNU Radio is dedicated to evolving a high-performance SDR.

All-digital radio transmitters and receivers present the possibility of advancing the capabilities of radio.

Crystal Radio

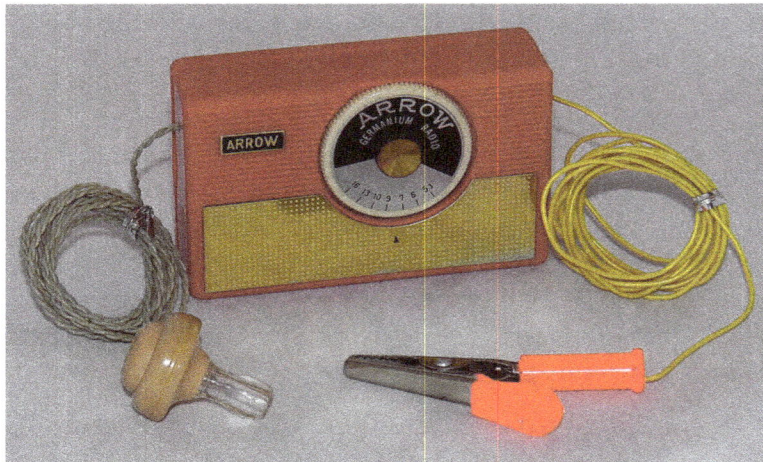

1970s-era crystal radio marketed to children. The earphone is on left. The antenna wire, right, has a clip to attach to metal objects such as a bedspring, which serve as an additional antenna to improve reception.

A crystal radio receiver, also called a crystal set or cat's whisker receiver, is a very simple radio receiver, popular in the early days of radio. It needs no other power source but that received solely from the power of radio waves received by a wire antenna. It gets its name from its most important component, known as a crystal detector, originally made from a piece of crystalline mineral such as galena. This component is now called a diode.

Crystal radios are the simplest type of radio receiver and can be made with a few inexpensive parts, such as a wire for an antenna, a coil of copper wire for adjustment, a capacitor, a crystal detector, and earphones. Crystal radios are distinct from ordinary radios as they are passive receivers, while other radios use a separate source of electric power such as a battery or the mains power to amplify the weak radio signal so as to make it louder. Thus, crystal sets produce rather weak sound and must be listened to with sensitive earphones, and can only receive stations within a limited range.

The rectifying property of crystals was discovered in 1874 by Karl Ferdinand Braun, and crystal detectors were developed and applied to radio receivers in 1904 by Jagadish Chandra Bose, G. W. Pickard and others. Crystal radios were the first widely used type of radio receiver, and the main type used during the wireless telegraphy era. Sold and homemade by the millions, the inexpensive and reliable crystal radio was a major driving force in the introduction of radio to the public, contributing to the development of radio as an entertainment medium with the beginning of radio broadcasting around 1920.

Around 1920, crystal sets were superseded by the first amplifying receivers, which used vacuum tubes, after which crystal sets became obsolete for commercial use. They continued to be built by hobbyists, youth groups, and the Boy Scouts however, as a way of learning about the technology of radio. Today they are still sold as educational devices, and there are groups of enthusiasts devoted to their construction.

Crystal radios receive amplitude modulated (AM) signals, and can be designed to receive almost any radio frequency band, but most receive the AM broadcast band. A few receive shortwave bands, but strong signals are required. The first crystal sets received wireless telegraphy signals broadcast by spark-gap transmitters at frequencies as low as 20 kHz.

History

A family listening to a crystal radio in the 1920s

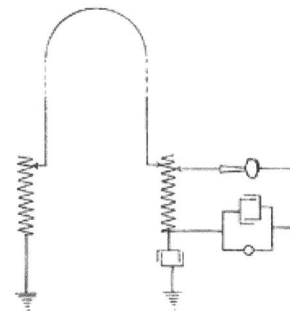

Greenleaf Whittier Pickard's US Patent 836,531 "Means for receiving intelligence communicated by electric waves" diagram

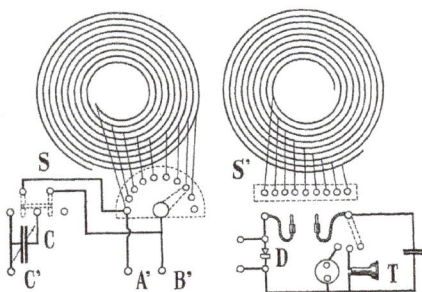

Radio receiver, Basel, Switzerland, 1914

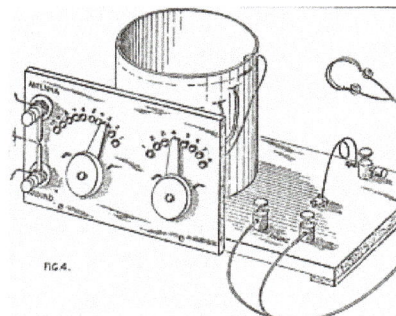

NBS Circular 120 Home Crystal Radio Project

Crystal radio was invented by a long, partly obscure chain of discoveries in the late 19th century that gradually evolved into more and more practical radio receivers in the early 20th century. The earliest practical use of crystal radio was to receive Morse code radio signals transmitted, from spark-gap transmitters, by early amateur radio experimenters. As electronics evolved, the ability to send voice signals by radio caused a technological explosion in the years around 1920 that evolved into today's radio broadcasting industry.

Crystal radio used as a backup receiver on a World War II Liberty ship

Early Years

Early radio telegraphy used spark gap and arc transmitters as well as high-frequency alternators running at radio frequencies. The coherer was the first means of detecting a radio signal. These, however, lacked the sensitivity to detect weak signals.

In the early 20th century, various researchers discovered that certain metallic minerals, such as galena, could be used to detect radio signals.

In 1901, Bose filed for a U.S. patent for "A Device for Detecting Electrical Disturbances" that mentioned the use of a galena crystal; this was granted in 1904, #755840. The device depended on the large variation of a semiconductor's conductance with temperature; today we would call his invention a bolometer. Bose's patent is frequently, but erroneously, cited as a type of rectifying detector. On August 30, 1906, Greenleaf Whittier Pickard filed a patent for a silicon crystal detector, which was granted on November 20, 1906. Pickard's detector was revolutionary in that he found that a fine pointed wire known as a "cat's whisker", in delicate contact with a mineral, produced the best semiconductor effect (that of rectification).

A crystal detector includes a crystal, a special thin wire that contacts the crystal, and the stand that holds those components in place. The most common crystal used is a small piece of galena; pyrite was also often used, as it was a more easily adjusted and stable mineral, and quite sufficient for urban signal strengths. Several other minerals also performed well as detectors. Another benefit of crystals was that they could demodulate amplitude modulated signals. This device brought radiotelephones and voice broadcast to a public audience. Crystal sets represented an inexpensive and technologically simple method of receiving these signals at a time when the embryonic radio broadcasting industry was beginning to grow.

1920s and 1930s

In 1922 the (then named) US Bureau of Standards released a publication entitled *Construction and Operation of a Simple Homemade Radio Receiving Outfit*. This article showed how almost any family having a member who was handy with simple tools could make a radio and tune into weather, crop prices, time, news and the opera. This design was significant in bringing radio to the general public. NBS followed that with a more selective two-circuit version, *Construction and Operation of a Two-Circuit Radio Receiving Equipment With Crystal Detector*, which was published the same year and is still frequently built by enthusiasts today.

In the beginning of the 20th century, radio had little commercial use, and radio experimentation was a hobby for many people. Some historians consider the autumn of 1920 to be the beginning of commercial radio broadcasting for entertainment purposes. Pittsburgh station KDKA, owned by Westinghouse, received its license from the United States Department of Commerce just in time to broadcast the Harding-Cox presidential election returns. In addition to reporting on special events, broadcasts to farmers of crop price reports were an important public service in the early days of radio.

In 1921, factory-made radios were very expensive. Since less-affluent families could not afford to own one, newspapers and magazines carried articles on how to build a crystal radio with common household items. To minimize the cost, many of the plans suggested winding the tuning coil on empty pasteboard containers such as oatmeal boxes, which became a common foundation for homemade radios.

Crystodyne

In early 1920s Russia, Oleg Losev was experimenting with applying voltage biases to various kinds of crystals for manufacture of radio detectors. The result was astonishing: with a zincite (zinc oxide) crystal he gained amplification. This was negative resistance phenomenon, decades before the development of the tunnel diode. After the first experiments, Losev built regenerative and super-heterodyne receivers, and even transmitters.

A crystodyne could be produced in primitive conditions; it can be made in a rural forge, unlike vacuum tubes and modern semiconductor devices. However, this discovery was not supported by authorities and soon forgotten; no device was produced in mass quantity beyond a few examples for research.

"Foxhole Radios"

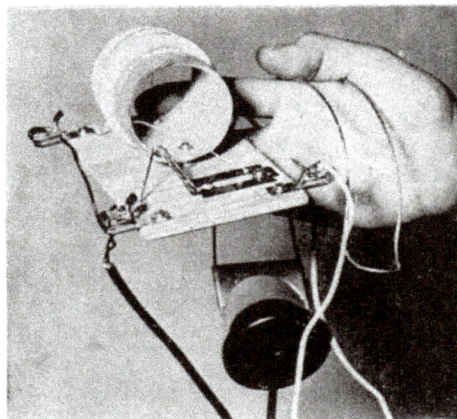

"Foxhole radio" used on the Italian Front in World War 2. It uses a pencil lead attached to a safety pin pressing against a razor blade for a detector.

In addition to mineral crystals, the oxide coatings of many metal surfaces act as semiconductors (detectors) capable of rectification. Crystal radios have been improvised using detectors made from rusty nails, corroded pennies, and many other common objects.

When Allied troops were halted near Anzio, Italy during the spring of 1944, powered personal radio receivers were strictly prohibited as the Germans had equipment that could detect the local oscillator signal of superheterodyne receivers. Crystal sets lack power driven local oscillators, hence they could not be detected. Some resourceful soldiers constructed "crystal" sets from discarded materials to listen to news and music. One type used a blue steel razor blade and a pencil lead for a detector. The lead point touching the semiconducting oxide coating (rust) on the blade formed a crude point-contact diode. By carefully adjusting the pencil lead on the surface of the blade, they could find sensitive spots, of iron oxide, capable of rectification. The lead of the pencil is made of graphite and clay and so it would inhibit further corrosion that would result if copper or iron wire was used in its place. Any further corrosion at the point of contact would ruin the diode effect found at that spot and further adjustment would be necessary. The sets were dubbed "foxhole radios" by the popular press, and they became part of the folklore of World War II.

In some German-occupied countries during WW2 there were widespread confiscations of radio sets from the civilian population. This led determined listeners to build their own "clandestine receivers" which frequently amounted to little more than a basic crystal set. However, anyone doing so risked imprisonment or even death if caught, and in most parts of Europe the signals from the BBC (or other allied stations) were not strong enough to be received on such a set.

Later Years

While it never regained the popularity and general use that it enjoyed at its beginnings, the crystal radio circuit is still used. The Boy Scouts have kept the construction of a radio set in their program since the 1920s. A large number of prefabricated novelty items and simple kits could be found through the 1950s and 1960s, and many children with an interest in electronics built one.

Building crystal radios was a craze in the 1920s, and again in the 1950s. Recently, hobbyists have started designing and building examples of the early instruments. Much effort goes into the visual appearance of these sets as well as their performance. Annual crystal radio 'DX' contests (long distance reception) and building contests allow these set owners to compete with each other and form a community of interest in the subject.

Design

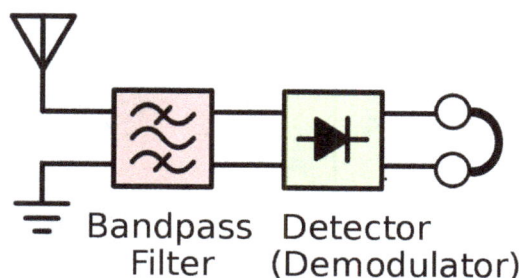

Block diagram of a crystal radio receiver

A crystal radio can be thought of as a radio receiver reduced to its essentials. It consists of at least these components:

- An antenna in which electric currents are induced by the radio waves.

- A resonant circuit (tuned circuit) which serves to select the frequency of the desired radio station out of all the radio signals received by the antenna. The tuned circuit consists of a coil of wire (called an inductor) and a capacitor connected together, so as to create a circuit that resonates at the frequency of the desired station, and hence "tune" in that station. One or both of the coil or capacitor is adjustable, allowing the circuit to be tuned to different frequencies. In some circuits a capacitor is not used, as the antenna also serves as the capacitor. The tuned circuit has a resonant frequency and allows radio waves at that frequency to pass to the detector, while rejecting waves at all other frequencies. Such a circuit is also known as a bandpass filter.

- A semiconductor crystal (detector) that demodulates the radio signal to get the audio signal (modulation). The crystal detector is a nonlinear impedance that functions as a square law detector. The detector's output is converted to sound by the earphone. Early sets used a cat's whisker detector, consisting of a fine wire touching the surface of a sample of crystalline mineral such as galena. It was this component that gave crystal sets their name.

- An earphone to convert the audio signal to sound waves so they can be heard. The low power produced by a crystal receiver is insufficient to power a loudspeaker, hence earphones are used.

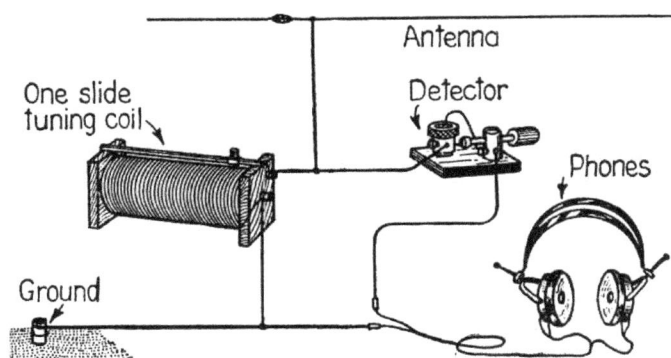

Pictorial diagram from 1922 showing the circuit of a crystal radio. This common circuit did not use a tuning capacitor, but used the capacitance of the antenna to form the tuned circuit with the coil. The detector might have been a piece of galena with a whisker wire in contact with it on a part of the crystal, making a diode contact

As a crystal radio has no power supply, the sound power produced by the earphone comes solely from the transmitter of the radio station being received, via the radio waves captured by the antenna. The power available to a receiving antenna decreases with the square of its distance from the radio transmitter. Even for a powerful commercial broadcasting station, if it is more than a few miles from the receiver the power received by the antenna is very small, typically measured in microwatts or nanowatts. In modern crystal sets, signals as weak as 50 picowatts at the antenna can be heard. Crystal radios can receive such weak signals without using amplification only due to the great sensitivity of human hearing, which can detect sounds with an intensity of only 10^{-16} W/cm^2. Therefore, crystal receivers have to be designed to convert the energy from the radio waves into sound waves as efficiently as possible. Even so, they are usually only able to receive stations within distances of about 25 miles for AM broadcast stations, although the radiotelegraphy signals used during the wireless telegraphy era could be received at hundreds of miles, and crystal receivers were even used for transoceanic communication during that period.

Commercial passive receiver development was abandoned with the advent of reliable vacuum tubes around 1920, and subsequent crystal radio research was primarily done by radio amateurs and hobbyists. Many different circuits have been used. The following sections discuss the parts of a crystal radio in greater detail.

Antenna

The antenna converts the energy in the electromagnetic radio waves to an alternating electric current in the antenna, which is connected to the tuning coil. Since in a crystal radio all the power comes from the antenna, it is important that the antenna collect as much power from the radio wave as possible. The larger an antenna, the more power it can intercept. Antennas of the type commonly used with crystal sets are most effective when their length is close to a multiple of a quarter-wavelength of the radio waves they are receiving. Since the length of the waves used with crystal radios is very long (AM broadcast band waves are 182-566 m or 597–1857 ft. long) the antenna is made as long as possible, from a long wire, in contrast to the whip antennas or ferrite loopstick antennas used in modern radios.

Serious crystal radio hobbyists use "inverted L" and "T" type antennas, consisting of hundreds of feet of wire suspended as high as possible between buildings or trees, with a feed wire attached in the center or at one end leading down to the receiver. However more often random lengths of wire dangling out windows are used. A popular practice in early days (particularly among apartment dwellers) was to use existing large metal objects, such as bedsprings, fire escapes, and barbed wire fences as antennas.

Ground

The wire antennas used with crystal receivers are monopole antennas which develop their output voltage with respect to ground. The receiver thus requires a connection to ground (the earth) as a return circuit for the current. The ground wire was attached to a radiator, water pipe, or a metal stake driven into the ground. In early days if an adequate ground connection could not be made a counterpoise was sometimes used. A good ground is more important for crystal sets than it is for powered receivers, as crystal sets are designed to have a low input impedance needed to transfer power efficiently from the antenna. A low resistance ground connection (preferably below 25 Ω) is necessary because any resistance in the ground reduces available power from the antenna. In contrast, modern receivers are voltage-driven devices, with high input impedance, hence little current flows in the antenna/ground circuit. Also, mains powered receivers are grounded adequately through their power cords, which are in turn attached to the earth by way of a well established ground.

Tuned Circuit

The tuned circuit, consisting of a coil and a capacitor connected together, acts as a resonator, similar to a tuning fork. Electric charge, induced in the antenna by the radio waves, flows rapidly back and forth between the plates of the capacitor through the coil. The circuit has a high impedance at the desired radio signal's frequency, but a low impedance at all other frequencies. Hence, signals at undesired frequencies pass through the tuned circuit to ground, while the desired frequency is instead passed on to the detector (diode) and stimulates the earpiece and is heard. The frequency of the station received is the resonant frequency f of the tuned circuit, determined by the capacitance C of the capacitor and the inductance L of the coil:

$$f = \frac{1}{2\pi\sqrt{LC}}$$

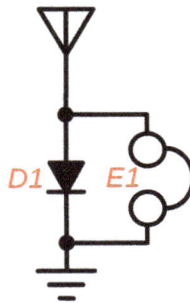

The earliest crystal receiver circuit did not have a tuned circuit

By varying either the inductor (L) or the capacitance (C), the circuit can be adjusted to different frequencies. In inexpensive sets, the inductor was made variable via a spring contact pressing against the windings that could slide along the coil, thereby introducing a larger or smaller number of turns of the coil into the circuit. Thus the inductance could be varied, "tuning" the circuit to the frequencies of different radio stations. Alternatively, a variable capacitor is used to tune the circuit. Some modern crystal sets use a ferrite core tuning coil, in which a ferrite magnetic core is moved into and out of the coil, thereby varying the inductance by changing the magnetic permeability (this eliminated the less reliable mechanical contact).

The antenna is an integral part of the tuned circuit and its reactance contributes to determining the circuit's resonant frequency. Antennas usually act as a capacitance, as antennas shorter than a quarter-wavelength have capacitive reactance. Many early crystal sets did not have a tuning capacitor, and relied instead on the capacitance inherent in the wire antenna (in addition to significant parasitic capacitance in the coil) to form the tuned circuit with the coil.

The earliest crystal receivers did not have a tuned circuit at all, and just consisted of a crystal detector connected between the antenna and ground, with an earphone across it. Since this circuit lacked any frequency-selective elements besides the broad resonance of the antenna, it had little ability to reject unwanted stations, so all stations within a wide band of frequencies were heard in the earphone (in practice the most powerful usually drowns out the others). It was used in the earliest days of radio, when only one or two stations were within a crystal set's limited range.

Impedance Matching

An important principle used in crystal radio design to transfer maximum power to the earphone is impedance matching. The maximum power is transferred from one part of a circuit to another when the impedance of one circuit is the complex conjugate of that of the other; this implies that the two circuits should have equal resistance. However, in crystal sets, the impedance of the antenna-ground system (around 10-200 ohms) is usually lower than the impedance of the receiver's tuned circuit (thousands of ohms at resonance), and also varies depending on the quality of the ground attachment, length of the antenna, and the frequency to which the receiver is tuned.

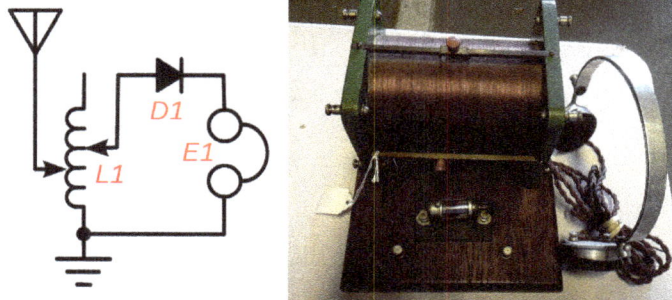

"Two slider" crystal radio circuit. and example from 1920s. The two sliding contacts on the coil allowed the impedance of the radio to be adjusted to match the antenna as the radio was tuned, resulting in stronger reception

Therefore, in improved receiver circuits, in order to match the antenna impedance to the receiver's impedance, the antenna was connected across only a portion of the tuning coil's turns. This made the tuning coil act as an impedance matching transformer (in an autotransformer connection) in addition to providing the tuning function. The antenna's low resistance was increased (transformed) by a factor equal to the square of the turns ratio (the ratio of the number of turns the antenna was connected to, to the total number of turns of the coil), to match the resistance across the tuned circuit. In the "two-slider" circuit, popular during the wireless era, both the antenna and the detector circuit were attached to the coil with sliding contacts, allowing (interactive) adjustment of both the resonant frequency and the turns ratio. Alternatively a multiposition switch was used to select taps on the coil. These controls were adjusted until the station sounded loudest in the earphone.

Problem of Selectivity

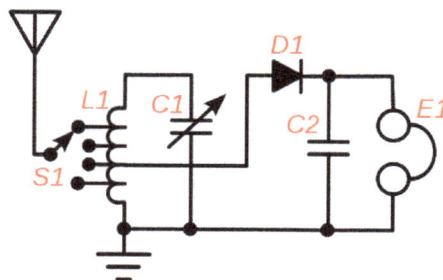

Direct-coupled circuit with impedance matching

One of the drawbacks of crystal sets is that they are vulnerable to interference from stations near in frequency to the desired station; that is to say, they have low selectivity. Often two or more stations are heard simultaneously. This is because the simple tuned circuit does not reject nearby signals well; it allows a wide band of frequencies to pass through, that is, it has a large bandwidth (low Q factor) compared to modern receivers.

The crystal detector worsened the problem, because it has relatively low resistance, thus it "loaded" the tuned circuit,, damping the oscillations (lowering the response), and reducing its Q. In many circuits, the selectivity was improved by connecting the detector and earphone circuit to a tap across only a fraction of the coil's turns. This reduced the impedance loading of the tuned circuit, as well as improving the impedance match with the detector.

Inductive Coupling

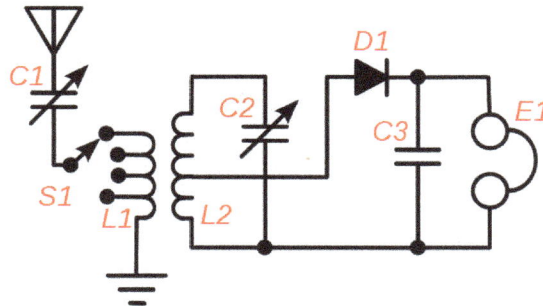

Inductively-coupled circuit with impedance matching. This type was used in most quality crystal receivers

In more sophisticated crystal receivers, the tuning coil is replaced with an adjustable air core antenna coupling transformer which improves the selectivity by a technique called *loose coupling*. This consists of two magnetically coupled coils of wire, one (the *primary*) attached to the antenna and ground and the other (the *secondary*) attached to the rest of the circuit. The current from the antenna creates an alternating magnetic field in the primary coil, which induced a current in the secondary coil which was then rectified and powered the earphone. Each of the coils functions as a tuned circuit; the primary coil resonated with the capacitance of the antenna (or sometimes another capacitor), and the secondary coil resonated with the tuning capacitor. Both the primary and secondary were tuned to the frequency of the station. The two circuits interacted to form a resonant transformer.

Reducing the *coupling* between the coils, by physically separating them so that less of the magnetic field of one intersects the other, reduces the mutual inductance, narrows the bandwidth, and results in much sharper, more selective tuning than that produced by a single tuned circuit. However, the looser coupling also reduced the power of the signal passed to the second circuit. The transformer was made with adjustable coupling, to allow the listener to experiment with various settings to gain the best reception.

One design common in early days, called a "loose coupler", consisted of a smaller secondary coil inside a larger primary coil. The smaller coil was mounted on a rack so it could be slid linearly in or out of the larger coil. If radio interference was encountered, the smaller coil would be slid further out of the larger, loosening the coupling, narrowing the bandwidth, and thereby rejecting the interfering signal.

The antenna coupling transformer also functioned as an impedance matching transformer, that allowed a better match of the antenna impedance to the rest of the circuit. One or both of the coils usually had several taps which could be selected with a switch, allowing adjustment of the number of turns of that transformer and hence the "turns ratio".

Coupling transformers were difficult to adjust, because the three adjustments, the tuning of the primary circuit, the tuning of the secondary circuit, and the coupling of the coils, were all interactive, and changing one affected the others.

Crystal Detector

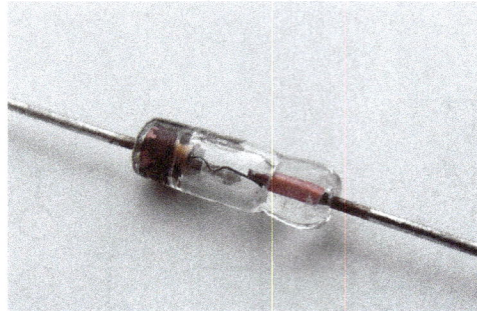

Germanium diode used in modern crystal radios (about 3 mm long)

The crystal detector demodulates the radio frequency signal, extracting the modulation (the audio signal which represents the sound waves) from the radio frequency carrier wave. In early receivers, the detector was a cat's whisker detector consisting of a fine metal wire, the "cat's whisker", on an adjustable arm that touched a pea-sized lump of semiconducting mineral. The point of contact between the wire and the crystal produced a diode effect. The cat's whisker detector was a crude Schottky diode that allowed current to flow better in one direction than in the opposite direction. Modern crystal sets use modern semiconductor diodes. The crystal functions as an envelope detector, rectifying the alternating current radio signal to a pulsing direct current, the peaks of which trace out the audio signal, so it can be converted to sound by the earphone, which is connected to the detector. The rectified current from the detector has radio frequency pulses from the carrier frequency in it, which are blocked by the high inductive reactance and do not pass well through the coils of early date earphones. Hence, a small capacitor called a bypass capacitor is often placed across the earphone terminals; its low reactance at radio frequency bypasses these pulses around the earphone to ground. In some sets the earphone cord had enough capacitance that this component could be omitted.

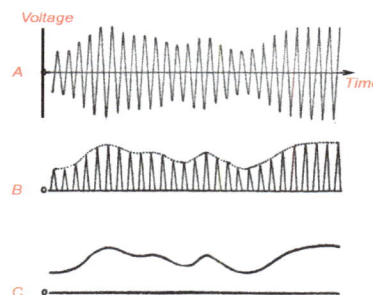

How the crystal detector works. *(A)* The amplitude modulated radio signal from the tuned circuit. The rapid oscillations are the radio frequency carrier wave. The audio signal (the sound) is contained in the slow variations (modulation) of the amplitude (hence the term amplitude modulation, AM) of the waves. This signal cannot be converted to sound by the earphone, because the audio excursions are the same on both sides of the axis, averaging out to zero, which would result in no net motion of the earphone's diaphragm. *(B)* The crystal conducts current better in one direction than the other, producing a signal whose amplitude does not average to zero but varies with the audio signal. *(C)* A bypass capacitor is used to remove the radio frequency carrier pulses, leaving the audio signal

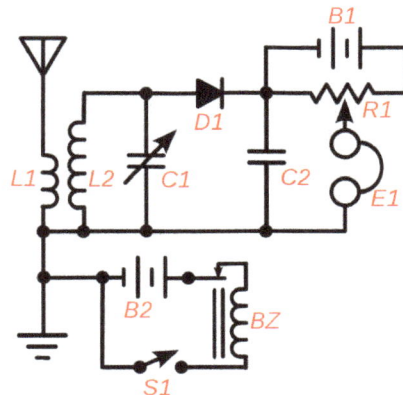

Circuit with detector bias battery to improve sensitivity and buzzer to adjust cat's whisker

In a cat's whisker detector only certain sites on the crystal surface functioned as rectifying junctions, and the device was very sensitive to the pressure of the crystal-wire contact, which could be disrupted by the slightest vibration. Therefore, a usable contact point had to be found by trial and error before each use. The operator dragged the wire across the crystal surface until a radio station or "static" sounds were heard in the earphones. Alternatively, some radios *(circuit, right)* used a battery-powered buzzer attached to the input circuit to adjust the detector. The spark at the buzzer's electrical contacts served as a weak source of static, so when the detector began working, the buzzing could be heard in the earphones. The buzzer was then turned off, and the radio tuned to the desired station.

Galena (lead sulfide) was probably the most common crystal used in cat's whisker detectors, but various other types of crystals were also used, the most common being iron pyrite (fool's gold, FeS_2), silicon, molybdenite (MoS_2), silicon carbide (carborundum, SiC), and a zincite-bornite (ZnO-Cu_5FeS_4) crystal-to-crystal junction trade-named *Perikon*. Crystal radios have also been improvised from a variety of common objects, such as blue steel razor blades and lead pencils, rusty needles, and pennies In these, a semiconducting layer of oxide or sulfide on the metal surface is usually responsible for the rectifying action.

In modern sets, a semiconductor diode is used for the detector, which is much more reliable than a cat's whisker detector and requires no adjustments. Germanium diodes (or sometimes Schottky diodes) are used instead of silicon diodes, because their lower forward voltage drop (roughly 0.3V compared to 0.6V) makes them more sensitive.

All semiconductor detectors function rather inefficiently in crystal receivers, because the low voltage input to the detector is too low to result in much difference between forward better conduction direction, and the reverse weaker conduction. To improve the sensitivity of some of the early crystal detectors, such as silicon carbide, a small forward bias voltage was applied across the detector by a battery and potentiometer. The bias moves the diode's operating point higher on the detection curve producing more signal voltage at the expense of less signal current (higher impedance). There is a limit to the benefit that this produces, depending on the other impedances of the radio. This improved sensitivity was caused by moving the DC operating point to a more desirable voltage-current operating point (impedance) on the junction's I-V curve. The battery did not power the radio, but only provided the biasing voltage which required little power.

Earphones

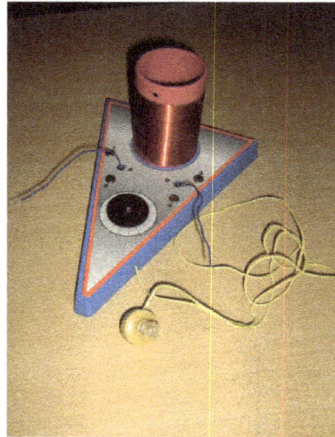

Modern crystal radio with piezoelectric earphone

The requirements for earphones used in crystal sets are different from earphones used with modern audio equipment. They have to be efficient at converting the electrical signal energy to sound waves, while most modern earphones sacrifice efficiency in order to gain high fidelity reproduction of the sound. In early homebuilt sets, the earphones were the most costly component.

The early earphones used with wireless-era crystal sets had moving iron drivers that worked in a way similar to the horn loudspeakers of the period; modern loudspeakers use a moving-coil principle. Each earpiece contained a permanent magnet about which was a coil of wire which formed a second electromagnet. Both magnetic poles were close to a steel diaphram of the speaker. When the audio signal from the radio was passed through the electromagnet's windings, current was caused to flow in the coil which created a varying magnetic field that augmented or diminished that due to the permanent magnet. This varied the force of attraction on the diaphragm, causing it to vibrate. The vibrations of the diaphragm push and pull on the air in front of it, creating sound waves. Standard headphones used in telephone work had a low impedance, often 75 Ω, and required more current than a crystal radio could supply. Therefore, the type used with crystal set radios (and other sensitive equipment) was wound with more turns of finer wire giving it a high impedance of 2000-8000 Ω.

Modern crystal sets use piezoelectric crystal earpieces, which are much more sensitive and also smaller. They consist of a piezoelectric crystal with electrodes attached to each side, glued to a light diaphragm. When the audio signal from the radio set is applied to the electrodes, it causes the crystal to vibrate, vibrating the diaphragm. Crystal earphones are designed as ear buds that plug directly into the ear canal of the wearer, coupling the sound more efficiently to the eardrum. Their resistance is much higher (typically megohms) so they do not greatly "load" the tuned circuit, allowing increased selectivity of the receiver. The piezoelectric earphone's higher resistance, in parallel with its capacitance of around 9 pF, creates a filter that allows the passage of low frequencies, but blocks the higher frequencies. In that case a bypass capacitor is not needed (although in practice a small one of around 0.68 to 1 nF is often used to help improve quality), but instead a 10-100 kΩ resistor must be added in parallel with the earphone's input.

Although the low power produced by crystal radios is typically insufficient to drive a loudspeaker, some homemade 1960s sets have used one, with an audio transformer to match the low impedance of the speaker to the circuit. Similarly, modern low-impedance (8 Ω) earphones cannot be used unmodified in crystal sets because the receiver does not produce enough current to drive them. They are sometimes used by adding an audio transformer to match their impedance with the higher impedance of the driving antenna circuit.

Use as a Power Source

A crystal radio tuned to a strong local transmitter can be used as a power source for a second amplified receiver of a distant station that cannot be heard without amplification.

There is a long history of unsuccessful attempts and unverified claims to recover the power in the carrier of the received signal itself. Traditional crystal sets use half-wave rectifiers. As AM signals have a modulation factor of only 30% by voltage at peaks, no more than 9% of received signal power ($P = U^2 / R$)) is actual audio information, and 91% is just rectified DC voltage. Given that the audio signal is unlikely to be at peak all the time, the ratio of energy is, in practice, even greater. Considerable effort was made to convert this DC voltage into sound energy. Some earlier attempts include a one-transistor amplifier in 1966. Sometimes efforts to recover this power are confused with other efforts to produce a more efficient detection. This history continues now with designs as elaborate as "inverted two-wave switching power unit".

Radio Transmitter Design

A radio transmitter is an electronic device which, when connected to an antenna, produces an electromagnetic signal such as in radio and television broadcasting, two way communications or radar. Heating devices, such as a microwave oven, although of similar design, are not usually called transmitters, in that they use the electromagnetic energy locally rather than transmitting it to another location.

Design Issues

A radio transmitter design has to meet certain requirements. These include the frequency of operation, the type of modulation, the stability and purity of the resulting signal, the efficiency of power use, and the power level required to meet the system design objectives. High-power transmitters may have additional constraints with respect to radiation safety, generation of X-rays, and protection from high voltages.

Typically a transmitter design includes generation of a carrier signal, which is normally sinusoidal, optionally one or more frequency multiplication stages, a modulator, a power amplifier, and a filter and matching network to connect to an antenna. A very simple transmitter might contain only a continuously running oscillator coupled to some antenna system. More elaborate transmitters allow better control over the modulation of the emitted signal and improve the stability of the transmitted frequency. For example, the Master Oscillator-Power Amplifier (MOPA) configuration inserts an amplifier stage between the oscillator and the antenna. This prevents changes in the loading presented by the antenna from altering the frequency of the oscillator.

Determining the Frequency

Fixed Frequency Systems

For a fixed frequency transmitter one commonly used method is to use a resonant quartz crystal in a Crystal oscillator to fix the frequency. Where the frequency has to be variable, several options can be used.

Variable Frequency Systems

- An array of crystals – used to enable a transmitter to be used on several different frequencies; rather than being a truly variable frequency system, it is a system which is fixed to several different frequencies (a subset of the above)

- Variable-frequency oscillator (VFO)

- Phase-locked loop frequency synthesiser

- Direct digital synthesis.

Frequency Multiplication

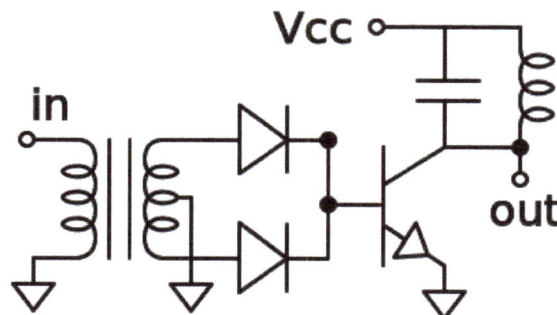

Frequency Doubler: A push-push frequency doubler. The output is tuned to two times the input frequency

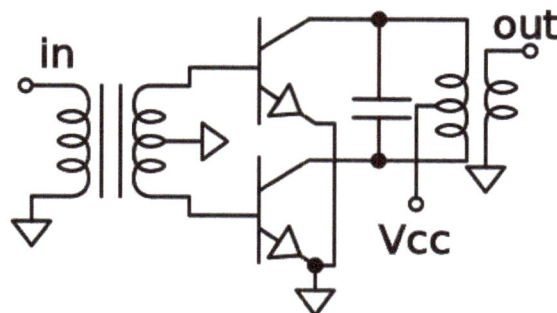

Frequency Tripler: A push-pull frequency tripler. The output is tuned to three times the input frequency

While modern frequency synthesizers can output a clean stable signal up through UHF, for many years, especially at higher frequencies, it was not practical to operate the oscillator at the final output frequency. For better frequency stability, it was common to multiply the frequency of the oscillator up to the final, required frequency. This was accommodated by allocating the short wave amateur and marine bands in harmonically related frequencies such as 3.5, 7, 14 and 28 MHz. Thus one crystal or VFO could cover several bands. In simple equipment this approach is still used occasionally.

If the output of an amplifier stage is simply tuned to a multiple of the frequency with which the stage is driven, the stage will give a large harmonic output. Many transmitters have used this simple approach successfully. However these more complex circuits will do a better job. In a push-push stage, the output will only contain *even* harmonics. This is because the currents which would generate the fundamental and the odd harmonics in this circuit are canceled by the second device. In a push-pull stage, the output will contain only *odd* harmonics because of the canceling effect.

Adding Modulation to the Signal

The task of a transmitter is to convey some form of information using a radio signal (carrier wave) which has been modulated to carry the intelligence. The RF generator in a microwave oven, electrosurgery, and induction heating are similar in design to transmitters, but usually not considered as such in that they do not intentionally produce a signal that will travel to a distant point. Such RF devices are required by law to operate in an ISM band where interference to radio communications will not occur. Where communications is the object, one or more of the following methods of incorporating the desired signal into the radio wave is used.

AM Modes

When the amplitude of a radio frequency wave is varied in amplitude in a manner which follows the modulating signal, usually voice, video or data, we have Amplitude modulation (AM).

Low Level and High Level

In low level modulation a small audio stage is used to modulate a low power stage. The output of this stage is then amplified using a linear RF amplifier. The great disadvantage of this system is that the amplifier chain is less efficient, because it has to be linear to preserve the modulation. Hence high efficiency class C amplifiers cannot be employed, unless a Doherty amplifier, EER (Envelope Elimination and Restoration) or other methods of predistortion or negative feedback are used. High level modulation uses class C amplifiers in a broadcast AM transmitter and only the final stage or final two stages are modulated, and all the earlier stages can be driven at a constant level. When modulation is applied to the plate of the final tube, a large audio amplifier is needed for the modulation stage, equal to 1/2 of the DC input power of the modulated stage. Traditionally the modulation is applied using a large audio transformer. However many different circuits have been used for high level AM modulation.

Types of AM Modulators

A wide range of different circuits have been used for AM. While it is perfectly possible to create good designs using solid-state electronics, valved (tube) circuits are shown here. In general, valves are able to easily yield RF powers far in excess of what can be achieved using solid state. Most high-power broadcast stations below 3 MHz use solid state circuits, but higher power stations above 3 MHz still use valves.

Plate AM Modulators

High level plate modulation consists of varying the voltage on the plate (anode) of the valve so that it swings from nearly zero to double the resting value. This will produce 100% modulation and can be done by inserting a transformer in series with the high voltage supply to the anode so that the vector sum of the two sources, (DC and audio) will be applied. A disadvantage is the size, weight and cost of the transformer as well as its limited audio frequency response, especially for very powerful transmitters.

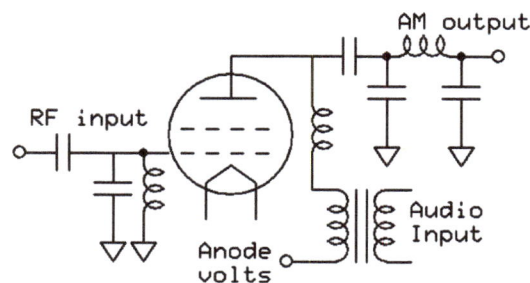

Anode modulation using a transformer. The valve anode sees the vector sum of anode volts and audio voltage.

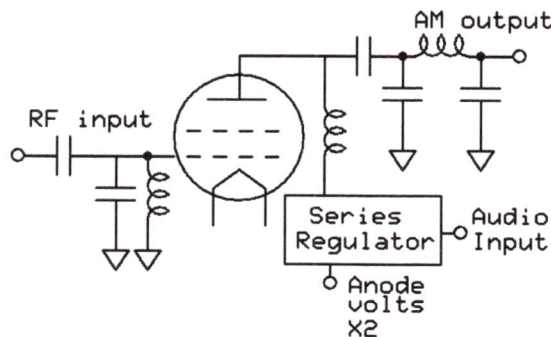

A series modulated stage. In modern transmitters the series regulator will use PWM switching for high efficiency. Historically the series regulator would have been a tube in analog mode.

Alternatively a series regulator can be inserted between the DC supply and the anode. The DC supply provides twice the normal voltage the anode sees. The regulator can allow none or all of the voltage to pass, or any intermediate value. The audio input operates the regulator in such a way as to produce the instantaneous anode voltage needed to reproduce the modulation envelope. An advantage of the series regulator is that it can set the anode voltage to any desired value. Thus the power output of the transmitter can be easily adjusted, allowing the use of Dynamic Carrier Control. The use of PDM switching regulators makes this system very efficient, whereas the original analog regulators were very inefficient and also non linear. Series PDM modulators are used in solid state transmitters also, but the circuits are somewhat more complex, using push pull or bridge circuits for the RF section.

These simplified diagrams omit such details as filament, screen and grid bias supplies, and the screen and cathode connections to RF ground.

Screen AM Modulators

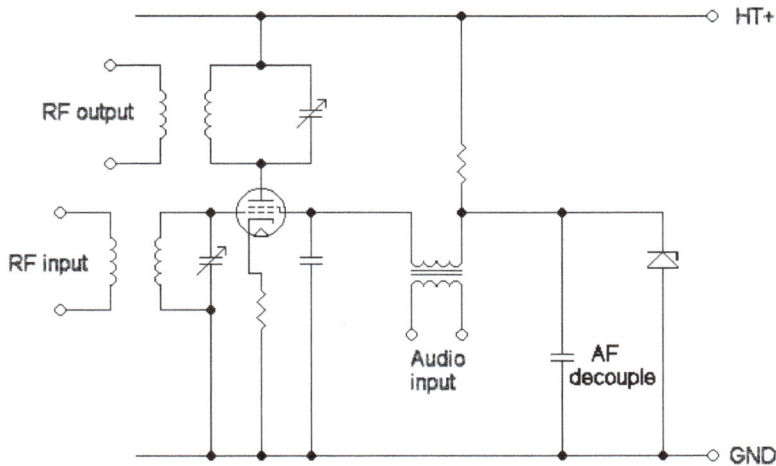

Screen AM modulator. Grid bias not shown

Under carrier conditions (no audio) the stage will be a simple RF amplifier where the screen voltage is set lower than normal to limit the RF output to about 25% of full power. When the stage is modulated the screen potential changes and so alters the gain of the stage. It takes much less audio power to modulate the screen, but final stage efficiency is only about 40%, compared to 80% with plate modulation. For this reason screen modulation was used only in low power transmitters and is now effectively obsolete.

AM Related Modes

Several derivatives of AM are in common use. These are

Single-sideband Modulation

SSB, or SSB-AM single-sideband full carrier modulation, is very similar to single-sideband suppressed carrier modulation (SSB-SC). It is used where it is necessary to receive the audio on an AM receiver, while using less bandwidth than with double sideband AM. Due to high distortion, it is seldom used. Either SSB-AM or SSB-SC are produced by the following methods.

Filter Method

Using a balanced mixer a double side band signal is generated, this is then passed through a very narrow bandpass filter to leave only one side-band. By convention it is normal to use the upper sideband (USB) in communication systems, except for amateur radio when the carrier frequency is below 10 MHz. There the lower side band (LSB) is normally used.

Phasing Method

The phasing method for the generation of single sideband signals uses a network which imposes a constant 90° phase shift on audio signals over the audio range of interest. This was difficult with analog methods but with DSP is very simple.

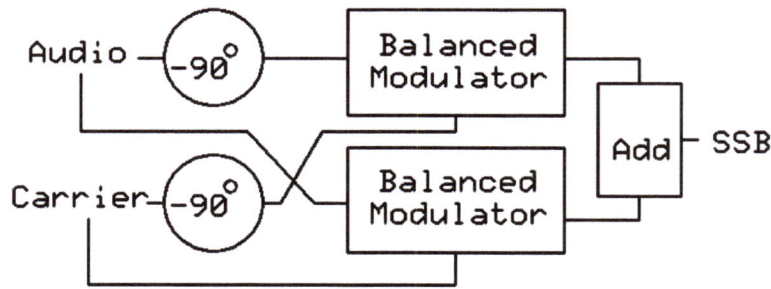

Phasing method of SSB generation

These audio outputs are each mixed in a linear balanced mixer with a carrier. The carrier drive for one of these mixers is also shifted by 90°. The outputs of these mixers are added in a linear circuit to give the SSB signal by phase cancellation of one of the sidebands. Connecting the 90° delayed signal from either the audio or the carrier (but not both) to the other mixer will reverse the sideband, so either USB or LSB is available with a simple DPDT switch.

Vestigial-sideband Modulation

Vestigial-sideband modulation (VSB, or VSB-AM) is a type of modulation system commonly used in analogue TV systems. It is normal AM which has been passed through a filter which reduces one of the sidebands. Typically, components of the lower sideband more than 0.75 MHz or 1.25 MHz below the carrier will be heavily attenuated.

Morse

Morse code is usually sent using on-off keying of an unmodulated carrier (Continuous wave). No special modulator is required.

This interrupted carrier may be analyzed as an AM-modulated carrier. On-off keying produces sidebands, as expected, but they are referred to as "key-clicks". Shaping circuits are used to turn the transmitter on and off smoothly instead of instantly in order to limit the bandwidth of these sidebands and reduce interference to adjacent channels.

FM Modes

Angle modulation is the proper term for modulation by changing the instantaneous frequency or phase of the carrier signal. True FM and phase modulation are the most commonly employed forms of analogue angle modulation.

Direct FM

Direct FM (true Frequency modulation) is where the frequency of an oscillator is altered to impose the modulation upon the carrier wave. This can be done by using a voltage-controlled capacitor (Varicap diode) in a crystal-controlled oscillator or frequency synthesiser. The frequency of the oscillator is then multiplied up using a frequency multiplier stage, or is translated upwards using a mixing stage, to the output frequency of the transmitter. The amount of modulation is referred to as the deviation, being the amount that the frequency of the carrier instantaneously deviates from the centre carrier frequency.

Indirect FM

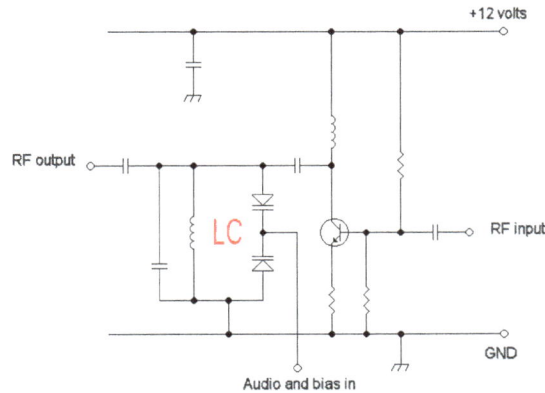

Indirect FM solid state circuit.

Indirect FM employs a varicap diode to impose a phase shift (which is voltage-controlled) in a tuned circuit that is fed with a plain carrier. This is termed phase modulation. In some indirect FM solid state circuits, an RF drive is applied to the base of a transistor. The tank circuit (LC), connected to the collector via a capacitor, contains a pair of varicap diodes. As the voltage applied to the varicaps is changed, the phase shift of the output will change.

Phase modulation is mathematically equivalent to direct Frequency modulation with a 6 dB/octave high-pass filter applied to the modulating signal. This high-pass effect can be exploited or compensated for using suitable frequency-shaping circuitry in the audio stages ahead of the modulator. For example, many FM systems will employ pre-emphasis and de-emphasis for noise reduction, in which case the high-pass equivalency of phase modulation automatically provides for the pre-emphasis. Phase modulators are typically only capable of relatively small amounts of deviation while remaining linear, but any frequency multiplier stages also multiply the deviation in proportion.

Digital Modes

Transmission of digital data is becoming more and more important. Digital information can be transmitted by AM and FM modulation, but often digital modulation consists of complex forms of modulation using aspects of both AM and FM. COFDM is used for DRM broadcasts. The transmitted signal consists of multiple carriers each modulated in both amplitude and phase. This allows very high bit rates and makes very efficient use of bandwidth. Digital or pulse methods also are used to transmit voice as in cell phones, or video as in terrestrial TV broadcasting. Early text messaging such as RTTY allowed the use of class C amplifiers, but modern digital modes require linear amplification.

Amplifying the Signal

Valves

Valves are electrically very robust, they can tolerate overloads which would destroy bipolar transistor systems in milliseconds. As a result, valved amplifiers may resist mistuning, lightning and power surges better. However, they require a heated cathode which consumes power and will fail in time due to loss of emission or heater burn out. The high voltages associated with valve circuits are dangerous to persons. For economic reasons, valves continue to be used for the final power amplifier for transmitters operating above 1.8 MHz and with powers above about 500 watts for amateur use and above about 10 Kw for broadcast use.

Solid State

Solid state devices, either discrete transistors or integrated circuits, are universally used for new transmitter designs up to a few hundred watts. The lower level stages of more powerful transmitters are also all solid state. Transistors can be used at all frequencies and power levels, but since the output of individual devices is limited, higher power transmitters must use many transistors in parallel, and the cost of the devices and the necessary combining networks can be excessive. As new transistor types become available and the price drops, solid state may eventually replace all valve amplifiers.

Linking the Transmitter to the Aerial

The majority of modern transmitting equipment is designed to operate with a resistive load fed via coaxial cable of a particular characteristic impedance, often 50 ohms. To connect the power stage of the transmitter to this coaxial cable transmission line a matching network is required. For solid state transmitters this is typically a broadband transformer which steps up the low impedance of the output devices to 50 ohms. A tube transmitter will contain a tuned output network, most commonly a PI network, that steps the load impedance which the tube requires down to 50 ohms. In each case the power producing devices will not transfer power efficiently if the network is detuned or badly designed or if the antenna presents other than 50 ohms at the transmitter output. Commonly an SWR meter and/or directional wattmeter are used to check the extent of the match between the aerial system and the transmitter via the transmission line (feeder). A directional wattmeter indicates forward power, reflected power, and often SWR as well. Each transmitter will specify a maximum allowable mismatch based on efficiency, distortion, and possible damage to the transmitter. Many transmitters have automatic circuits to reduce power or shut down if this value is exceeded.

Transmitters feeding a balanced transmission line will need a balun. This transforms the single ended output of the transmitter to a higher impedance balanced output. High power short wave transmission systems typically use 300 ohm balanced lines between the transmitter and antenna. Amateurs often use 300-450 ohm balanced antenna feeders.

EMC Matters

Many devices depend on the transmission and reception of radio waves for their operation. The possibility for mutual interference is great. Many devices not intended to transmit signals may do so. For instance a dielectric heater might contain a 2000 watt 27 MHz source within it. If the machine operates as intended then none of this RF power will leak out. However, if due to poor design or maintenance it allows RF to leak out, it will become a transmitter or unintentional radiator.

RF Leakage and Shielding

All equipment using RF electronics should be inside a screened conductive box and all connections in or out of the box should be filtered to avoid the passage of radio signals. A common and effective method of doing so for wires carrying DC supplies, 50/60 Hz AC connections, audio and control signals is to use a feedthrough capacitor, whose job is to short circuit any RF on the wire to ground. The use of ferrite beads is also common.

If an intentional transmitter produces interference, then it should be run into a dummy load; this is a resistor in a screened box or can which will allow the transmitter to generate radio signals without sending them to the antenna. If the transmitter continues to cause interference during this test then a path exists by which RF power is leaking out of the equipment and this can be due to bad shielding. Such leakage is most likely to occur on homemade equipment or equipment that has been modified or had covers removed. RF leakage from microwave ovens, while rare, may occur due to defective door seals, and may be a health hazard.

Spurious Emissions

Early in the development of radio technology it was recognized that the signals emitted by transmitters had to be 'pure'. Spark-gap transmitters were outlawed once better technology was available as they give an output which is very wide in terms of frequency. The term spurious emissions refers to any signal which comes out of a transmitter other than the wanted signal. In modern equipment there are three main types of spurious emissions: harmonics, out of band mixer products which are not fully suppressed and leakage from the local oscillator and other systems within the transmitter.

Harmonics

These are multiples of the operation frequency of the transmitter, they can be generated in any stage of the transmitter which is not perfectly linear and must be removed by filtering.

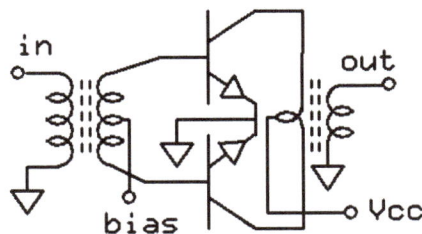

This push pull wide band amplifier uses ferrite core transformers for matching and coupling. The two NPN transistors can be biased to class A, AB or C, and will still have very weak harmonics at even multiples of the design frequency. The odd harmonics will be stronger, but still manageable. Class C will have the most harmonics.

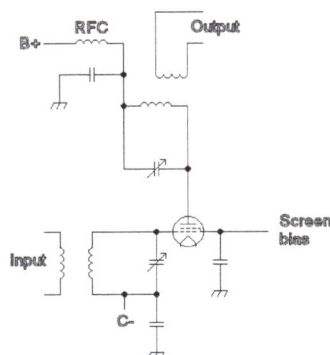

This single ended amplifier uses a narrowly tuned anode circuit to reduce harmonics when operating class AB or C.

Avoiding Harmonic Generation

The difficulty of removing harmonics from an amplifier will depend on the design. A push-pull amplifier will have fewer harmonics than a single ended circuit. A class A amplifier will have very few harmonics, class AB or B more, and class C the most. In the typical class C amplifier, the resonant tank circuit will remove most of the harmonics, but in either of these examples, a low pass filter will likely be needed following the amplifier.

Removal of Harmonics with Filters

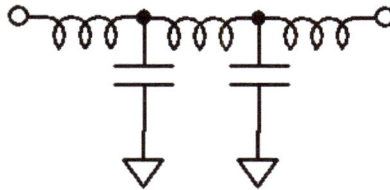

A simple low pass filter suitable for harmonic reduction.

In addition to the good design of the amplifier stages, the transmitter's output should be filtered with a low pass filter to reduce the level of the harmonics. Typically the input and output are interchangeable and match to 50 ohms. Inductance and capacity values will vary with frequency. Many transmitters switch in a suitable filter for the frequency band being used. The filter will pass the desired frequency and reduce all harmonics to acceptable levels.

The harmonic output of a transmitter is best checked using an RF spectrum analyzer or by tuning a receiver to the various harmonics. If a harmonic falls on a frequency being used by another communications service then this spurious emission can prevent an important signal from being received. Sometimes additional filtering is used to protect a sensitive range of frequencies, for example, frequencies used by aircraft or services involved with protection of life and property. Even if a harmonic is within the legally allowed limits, the harmonic should be further reduced.

Oscillators and Mix Products

When mixing signals to produce a desired output frequency, the choice of Intermediate frequency and local oscillator is important. If poorly chosen, a spurious output can be generated. For example, if 50 MHz is mixed with 94 MHz to produce an output on 144 MHz, the third harmonic of the 50 MHz may appear in the output. This problem is similar to the Image response problem which exists in receivers.

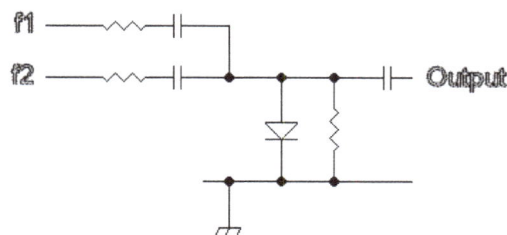

Simple but poor mixer. A diode is shown but any non-linear device can be used.

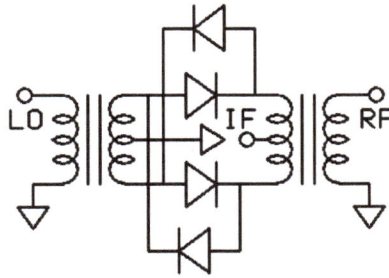

A double balanced mixer using matched diodes. It is also possible to use active devices such as transistors or valves.

One method of reducing the potential for this transmitter defect is the use of balanced and double balanced mixers. A simple mixer will pass both of the input frequencies and all of their harmonics along with the sum and difference frequencies. If the simple mixer is replaced with a balanced mixer then the number of possible products is reduced. If the frequency mixer has fewer outputs the task of making sure that the final output is *clean* will be simpler.

Instability and Parasitics

If a stage in a transmitter is unstable and is able to oscillate then it can start to generate RF at either a frequency close to the operating frequency or at a very different frequency. One good sign that it is occurring is if an RF stage has a power output even without being driven by an exciting stage. Output power should increase smoothly as input power is increased, although with Class C, there will be a noticeable threshold effect. Various circuits are used for parasitic suppression in a good design. Proper neutralization is also important.

Control and Protection

Yaesu FT-817 Transceiver controls One button and two knobs allow control of 52 separate parameters.

The simplest transmitters such as RFID devices require no external controls. Simple tracking transmitters may have only an on-off switch. Many transmitters must have circuits that allow them to be turned on and off and the power output and frequency adjusted or modulation levels adjusted. Many modern multi-featured transmitters allow the adjustment of many different parameters. Usually these are under microprocessor control via multilevel menus, thus reducing the required number of physical knobs. Often a display screen provides feedback to the operator to assist in adjustments. The user friendliness of this interface will often be one of the main factors in a successful design.

Microprocessor controlled transmitters also may include software to prevent off frequency or other illegal operation. Transmitters using significant power or expensive components must also have protection circuits which prevent such things as overload, overheating or other abuse of the circuits. Overload circuits may include mechanical relays, or electronic circuits. Simple fuses may be included to protect expensive components. Arc detectors may shut off the transmitter when sparks or fires occur.

Protection features must also prevent the human operator and the public from encountering the high voltages and power which exist inside the transmitter. Tube transmitters typically use DC voltages between 600 and 30,000 volts, which are deadly if contacted. Radio frequency power above about 10 watts can cause burning of human tissue through contact and higher power can actually cook human flesh without contact. Metal shielding is required to isolate these dangers. Properly designed transmitters have doors or panels which are interlocked, so that open doors activate switches which do not allow the transmitter to be turned on when the dangerous areas are exposed. In addition, either resistors which bleed off the high voltages or shorting relays are employed to insure that capacitors do not retain a dangerous charge after turn off.

With large high power transmitters, the protective circuits can comprise a significant fraction of the total design complexity and cost.

Power Supplies

Some RFID devices take power from an external source when it interrogates the device, but most transmitters either have self-contained batteries, or are mobile systems which typically operate directly from the 12 volt vehicle battery. Larger fixed transmitters will require power from the mains. The voltages used by a transmitter will be AC and DC of many different values. Either AC transformers or DC power supplies are required to provide the values of voltage and current needed to operate the various circuits. Some of these voltages will need to be regulated. Thus a significant part of the total design will consist of power supplies. Power supplies will be integrated into the control and protection systems of the transmitter, which will turn them on in the proper sequence and protect them from overloads. Often rather complicated logic systems will be required for these functions.

Radio Repeater

A radio repeater is a combination of a radio receiver and a radio transmitter that receives a weak or low-level signal and retransmits it at a higher level or higher power, so that two-way radio signals can cover longer distances without degradation. This section refers to professional, commercial, and government radio systems. A separate article exists for Amateur radio repeaters.

In dispatching, amateur radio, and emergency services communications, repeaters are used extensively to relay radio signals across a wider area. With most emergency (and some other) dispatching systems, the repeater is synonymous with the base station, which performs both functions. This includes police, fire brigade, ambulance, taxicab, tow truck, and other services. The General Mobile Radio Service in the United States and UHF CB service in Australia also use repeaters in much the same fashion as amateur radio operators do.

A continuous-duty, rack-mount iDEN digital trunked system repeater at a cell site.

Full Duplex Operation

Motorola MOTOTRBO Repeater DR3000 with duplexer mounted in Flightcase, 100% Duty cycle up to 40 W output

A repeater is an automatic radio-relay station, usually located on a mountain top, tall building, or radio tower. It allows communication between two or more bases, mobile or portable stations that are unable to communicate directly with each other due to distance or obstructions between them.

The repeater receives on one radio frequency (the "input" frequency), demodulates the signal, and simultaneously re-transmits the information on its "output" frequency. All stations using the repeater transmit on the repeater's input frequency and receive on its output frequency. Since the repeater is usually located at an elevation higher than the other radios using it, their range is greatly extended.

Because the transmitter and receiver are on at the same time, isolation must exist to keep the repeater's own transmitter from degrading the repeater receiver. If the repeater transmitter and receiver are not isolated well, the repeater's own transmitter *desensitizes* the repeater receiver. The problem is similar to being at a rock concert and not being able to hear the weak signal of a conversation over the much stronger signal of the band.

In general, isolating the receiver from the transmitter is made easier by maximizing, as much as possible, the separation between input and output frequencies.

Frequency Separation: Input to Output

There is no set rule about spacing of input and output frequencies for all radio repeaters. Any spacing where the designer can get sufficient isolation between receiver and transmitter will work.

The narrowest documented spacing found in preparing this article is a system dismantled in the 1980s. One channel used on the Riverside County, California Sheriff's Department system, which was replaced by an 800 MHz trunked system, used 158.760 MHz as input and 159.090 MHz as output, a spacing of 330 kHz. The station callsign was KMH971. It is unusual to see systems with input and output spaced this closely. It is believed working systems have been spaced as close as 175 kHz. There is currently a narrow spacing in use in Alamogordo, NM used for the city EMS dispatching. Input is 155.715 MHz and output is 155.385 MHz which is also a spacing of 330 kHz.

In some countries, under some radio services, there are agreed-on conventions or separations that are required by the system license. In the case of input and output frequencies in the United States, for example:

- Amateur repeaters in the 144–148 MHz band usually use a 600 kHz (0.6 MHz) separation.

- Systems in the 450–470 MHz band use a 5 MHz separation with the input on the higher frequency. Example: input is 456.900 MHz; output is 451.900 MHz.

- Amateur repeaters in the 420–450 MHz band use a 5 MHz separation with the input being either on the higher or lower portion of the 440–450 segment of the band (the standard changes regionally). Example: output is 444.400 MHz; input is 449.400 MHz.

- Systems in the 806–869 MHz band use a 45 MHz separation with the input on the lower frequency. Example: input is 810.1875 MHz; output is 855.1875 MHz.

- Amateur repeaters in the 902–928 MHz band use a 25 MHz separation, with the inputs on the lower frequency.

- Military systems are suggested to use no less than a 10 MHz spacing.

These are just a few examples. There are many other separations or spacings between input and output frequencies in operational systems.

Same Band Frequencies

Same band repeaters operate with input and output frequencies in the same frequency band. For example, in US two-way radio, 30–50 MHz is one band and 150–174 MHz is another. A repeater with an input of 33.980 MHz and an output of 46.140 MHz is a same band repeater.

In same band repeaters, a central design problem is keeping the repeater's own transmitter from interfering with the receiver. Reducing the coupling between transmitter and input frequency receiver is called *isolation*.

Duplexer System

In same-band repeaters, isolation between transmitter and receiver can be created by using a single antenna and a device called a *duplexer*. The device is a tuned filter connected to the antenna. In this example, consider a type of device called a *band-pass duplexer*. It allows, or passes, a band, (or a narrow range,) of frequencies.

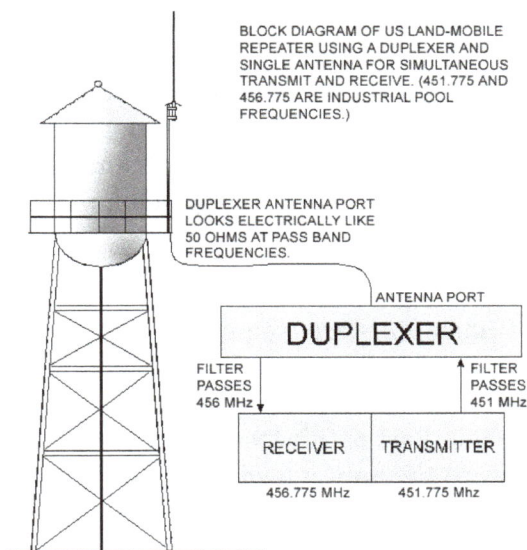

BLOCK DIAGRAM OF US LAND-MOBILE REPEATER USING A DUPLEXER AND SINGLE ANTENNA FOR SIMULTANEOUS TRANSMIT AND RECEIVE. (451.775 AND 456.775 ARE INDUSTRIAL POOL FREQUENCIES.)

DUPLEXER ANTENNA PORT LOOKS ELECTRICALLY LIKE 50 OHMS AT PASS BAND FREQUENCIES.

ANTENNA PORT

DUPLEXER

FILTER PASSES 456 MHz

FILTER PASSES 451 MHz

RECEIVER | TRANSMITTER

456.775 MHz | 451.775 Mhz

There are two legs to the duplexer filter, one is tuned to pass the input frequency, the other is tuned to pass the output frequency. Both legs of the filter are coupled to the antenna. The repeater receiver is connected to the input leg while the transmitter is connected to the output leg. If the right specifications are chosen, the duplexer has a narrow-enough filter to prevent the repeater's receiver from being overloaded by its own transmitter. By virtue of the transmitter and receiver being on different frequencies, they can operate at the same time on a single antenna.

Any anomaly or fault with the antenna or antenna feed cable will reflect transmitter power back into the receiver, possibly causing the receiver to be overloaded. The reflected power will quickly exceed the duplexer's filtering ability.

Combining System

There is often not enough tower space to accommodate a separate antenna for each repeater at crowded equipment sites. In same-band repeaters at engineered, shared equipment sites, repeaters can be connected to shared antenna systems. These are common in trunked systems, where up to 29 repeaters for a single trunked system may be located at the same site. (Some architectures such as iDEN sites may have more than 29 repeaters.)

In a shared system, a receive antenna is usually located at the top of the antenna tower. Putting the receive antenna at the top helps to capture weaker received signals than if the receive antenna were lower of the two. By splitting the received signal from the antenna, many receivers can work satisfactorily from a single antenna. Devices called *receiver multicouplers* split the signal from the antenna into many receiver connections. The multicoupler amplifies the signals reaching the antenna, then feeds them to several receivers, attempting to make up for losses in the power dividers (or splitters). These operate similarly to a cable TV splitter but must be built to higher quality standards so they work in environments where strong interfering signals are present.

On the transmitter side, a second transmit antenna is installed somewhere below the receive antenna. There is an electrical relationship defined by the distance between transmit and receive antennas. A desirable null exists if the transmit antenna is located exactly below the receive antenna beyond a minimum distance. Almost the same isolation as a low-grade duplexer (about −60 decibels) can be accomplished by installing the transmit antenna below, and along the centerline of, the receive antenna. Several transmitters can be connected to the same antenna using filters called *combiners*. Transmitters usually have directional devices installed along with the filters that block any reflected power in the event the antenna malfunctions. The antenna must have a power rating that will handle the sum of energy of all connected transmitters at the same time.

Transmitter combining systems are lossy. As a rule of thumb, each leg of the combiner has a 50% (3 decibel) power loss. If two transmitters are connected to a single antenna through a combiner, half of their power will reach the combiner output. (This assumes everything is working properly.) If four transmitters are coupled to one antenna, a quarter of each transmitter's power will reach the output of the combining circuit. Part of this loss can be made up with increased antenna gain. Fifty watts of transmitter power to the antenna will make a received signal strength at a distant mobile radio that is almost identical to 100 watts.

In trunked systems with many channels, a site design may include several transmit antennas to reduce combining network losses. For example, a six-channel trunked system may have two transmit antennas with three transmitters connected to each of the two transmit antennas. Because small variations affect every antenna, each antenna will have a slightly different directional pattern. Each antenna will interact with the tower and other nearby antennas differently. If one were to measure received signal levels, this would cause a variation among channels on a single trunked system. Variations in signal strength among channels on one trunked system can also be caused by:

- failed parts in the combiner,

- characteristics of the design,

- loose connectors,

- bad cables,

- mistuned filters, or;

- incorrectly installed components.

Cross-band Frequencies

Consider the possibility that amateurs are likely to have been the first to experiment with, and to use, cross band repeaters.

Modern

Cross-band repeaters are sometimes a part of government trunked radio systems. If one community is on a trunked system and the neighboring community is on a conventional system, a talk group or agency-fleet-subfleet may be designated to communicate with the other community. In an example where the community is on 153.755 MHz, transmitting on the trunked system talk group would repeat on 153.755 MHz. Signals received by a base station on 153.755 MHz would go over the trunked system on an assigned talk group.

In conventional government systems, cross band repeaters are sometimes used to connect two agencies who use radio systems on different bands. For example, a fire department in Colorado was on a 46 MHz channel while a police department was on a 154 MHz channel, they built a crossband repeater to allow communication between the two agencies.

If one of the systems is simplex, the repeater must have logic preventing transmitter keying in both directions at the same time. Voting comparators with a transmitter keying matrix are sometimes used to connect incompatible base stations.

Historic

In looking at records of old systems, examples of cross-band commercial systems were found in every US radio service where regulations allowed them. In California, specific systems using cross-band repeaters have existed at least since the 1960s. Historic examples of cross-band systems include:

- Solano County Fire, (former Fire Radio Service): 46.240 input; 154.340 output. This system was dismantled in the 1980s and is now a same band repeater.

- Mid-Valley Fire District, Fresno, (former Fire Radio Service): 46.140 input; 154.445 output. This system was dismantled in the 1980s and is now a same band repeater.

- Santa Clara County Department of Parks and Recreation, (former Forestry Conservation Radio Service): 44.840 MHz input; 151.445 MHz output. This system was dismantled in the 1980s and is now a same band repeater.

- State of California, Governor's Office of Emergency Services, Fire, (former Fire Radio Service): 33.980 MHz input; 154.160 MHz output.

In commercial systems, manufacturers stopped making cross band mobile radio equipment with acceptable specifications for public safety systems in the early 1980s. At the time, some systems were dismantled because new radio equipment was not available.

As Links

For decades, cross-band repeaters have been used as fixed links. The links can be used for remote control of base stations at distant sites or to send audio from a diversity (voting) receiver site back to the diversity combining system (voting comparator). Some legacy links occur in the US 150–170 MHz band. US Federal Communications Commission rule changes did not allow 150 MHz links after the 1970s. Newer links are more often seen on 72–76 MHz (Mid-band), 450–470 MHz interstitial channels, or 900 MHz links. These links, known as *fixed* stations in US licensing, typically connect an equipment site with a dispatching office.

Vehicular Repeaters

Modern amateur radios sometimes include cross-band repeat capability native to the radio transceiver.

In commercial systems, cross-band repeaters are sometimes used in vehicular repeaters. For example, a 150 MHz hand held may communicate to a vehicle-mounted low-power transceiver. The low-power radio repeats transmissions from the portable over the vehicle's high power mobile radio, which has a much longer range. In these systems, the hand-held works so long as it is within range of the low power mobile repeater. The mobile radio is usually on a different band than the hand-held to reduce the chances of the mobile radio transmitter interfering with the transmission from the hand-held to the vehicle.

- Motorola, for example, marketed a vehicular repeater system called PAC*RT. It was available for use with 150 MHz or 450 MHz hand-helds and interfaced with some Motorola mobile radios.

- In the 1980s, General Electric Mobile Radio had a 463 MHz emergency medical services radio that featured a 453 MHz vehicular repeater link to a hand-held.

There is a difficult engineering problem with these systems. If you get two vehicle radios at the same location, some protocol has to be established so that one portable transmitting doesn't activate two or more mobile radio transmitters. Motorola uses a hierarchy system with PAC*RT, each repeater transmits a tone when it is turned on, so the last one on site that turns on is the one that gets used. This is so several of them are not on at once.

Vehicular repeaters are complex but can be less expensive than designing a system that covers a large area and works with the weak signal levels of hand-held radios. Some models of radio signals suggest that the transmitters of hand-held radios create received signals at the base station one to two orders of magnitude (10 to 20 decibels or 10 to 100 times) weaker than a mobile radio with a similar transmitter output power.

Siting as Part of System Design

Radio repeaters are typically placed in locations which maximize their effectiveness for their intended purpose:

- "Low-level" repeaters are used for local communications, and are placed at low altitude to reduce interference with other users of the same radio frequencies. Low-level systems are used for areas as large as an entire city, or as small as a single building.

- "High-level" repeaters are placed on tall towers or mountaintops to maximize their area of coverage. With these systems, users with low-powered radios (such as hand-held "walkie-talkies") can communicate with each other over many miles.

Amateur Radio Repeater

An amateur radio repeater is an electronic device that receives a weak or low-level amateur radio signal and retransmits it at a higher level or higher power, so that the signal can cover longer distances without degradation. Many repeaters are located on hilltops or on tall buildings as the higher location increases their coverage area, sometimes referred to as the radio horizon, or "footprint". Amateur radio repeaters are similar in concept to those used by public safety entities (police, fire department, etc.), businesses, government, military, and more. Amateur radio repeaters may even use commercially packaged repeater systems that have been adjusted to operate within amateur radio frequency bands, but more often amateur repeaters are assembled from receivers, transmitters, controllers, power supplies, antennas, and other components, from various sources.

An amateur radio repeater system consisting of a 70cm repeater and a 2-meter digipeater and iGate.

Coaxial cavity RF filter at 2 meter repeater

Introduction

In amateur radio, repeaters are typically maintained by individual hobbyists or local groups of amateur radio operators. Many repeaters are provided openly to other amateur radio operators and typically not used as a remote base station by a single user or group. In some areas multiple repeaters are linked together to form a wide-coverage network, such as the linked system provided by the Independent Repeater Association which covers most of western Michigan, or the Western Intertie Network System ("WINsystem") that now covers a great deal of California, and is in 17 other states, including Hawaii, along with parts of four other countries, Australia, Canada, Great Britain and Japan.

Frequencies

Repeaters are found mainly in the VHF six meters (50—54 MHz), two meter (144—148 MHz), 1.25-meter band (1 1/4 meter) (220—224 MHz) and the UHF 70 centimeter (420—450 MHz) bands, but can be used on almost any frequency pair above 28 MHz. Recently, 33 centimeters (902—928 MHz) and 23 centimeters (1,240—1,300 MHz) are also being used for repeaters. Note that different countries have different rules; for example, in the United States, the two meter band is 144—148 MHz, while in the United Kingdom (and most of Europe) it is 144—147 MHz.

Repeater frequency sets are known as "repeater pairs", and in the ham radio community most follow *ad hoc* standards for the difference between the two frequencies, commonly called the *offset*. In the USA two-meter band, the standard offset is 600 kHz (0.6 MHz), but sometimes unusual offsets, referred to as *oddball splits*, are used. The actual frequency pair used is assigned by a local frequency coordinating council.

In the days of crystal-controlled radios, these pairs were identified by the last portion of the transmit *(Input)* frequency followed by the last portion of the receive *(Output)* frequency that the ham would put into the radio. Thus "three-four nine-four" (34/94) meant that hams would transmit on 146.34 MHz and listen on 146.94 MHz (while the repeater would do the opposite, listening on 146.34 and transmitting on 146.94). In areas with many repeaters, "reverse splits" were common (i.e., 94/34), to prevent interference between systems.

Since the late 1970s, the use of synthesized, microprocessor-controlled radios, and widespread adoption of standard frequency splits have changed the way repeater pairs are described. In 1980, a ham might have been told that a repeater was on "22/82"—today they will most often be told "682 down". The 6 refers to the last digit of 146 MHz, so that the display will read "146.82" (the output frequency), and the radio is set to transmit "down" 600 kHz on 146.22 MHz. Another way of describing a repeater frequency pair is to give the repeater's output frequency, along with the direction of offset ("+" or "plus" for an input frequency above the output frequency, "–" or "minus" for a lower frequency) with the assumption that the repeater uses the standard offset for the band in question. For instance, a 2-meter repeater might be described as "147.34 with a plus offset", meaning that the repeater transmits on 147.34 MHz and receives 600 kHz above the output frequency.

Services

Services provided by a repeater may include an autopatch connection to a POTS/PSTN telephone line to allow users to make telephone calls from their keypad-equipped radios. These advanced services may be limited to members of the group or club that maintains the repeater. Many amateur radio repeaters typically have a tone access control (CTCSS, CG or PL tone) implemented to prevent them from being keyed-up (operated) accidentally by interference from other radio signals. A few use a digital code system called *DCS, DCG* or *DPL* (a Motorola trademark). In the UK most repeaters also respond to a short burst of 1750 Hz tone to open the repeater.

In many communities, a repeater has become a major on-the-air gathering spot for the local amateur radio community, especially during "drive time" (the morning or afternoon commuting time). In the evenings local public service nets may be heard on these systems and many repeaters are used by weather spotters. In an emergency or a disaster a repeater can sometimes help to provide needed communications between areas that could not otherwise communicate. Until cellular telephones became popular, it was common for community repeaters to have "drive time" monitoring stations so that mobile amateurs could call in traffic accidents via the repeater to the monitoring station who could relay it to the local police agencies via telephone. Systems with autopatches frequently had (and still have) most of the public safety agencies numbers programmed as speed-dial numbers.

US Repeater Coordination

Repeater coordination is not required by the Federal Communications Commission, nor does the FCC regulate, certify or otherwise regulate frequency coordination for the Amateur Radio Bands.

Amateur Radio Repeater Coordinators or coordination groups are all volunteers and have no legal authority to assume jurisdictional or regional control in any area where the Federal Communications Commission regulates the Amateur Radio Service. The United States Code of Federal Regulations Title 47 CFR, Part 97, which are the laws in which the Amateur Radio Service is regulated clearly states the definition of Frequency Coordinator.

The purpose of coordinating a repeater or frequency is reduce harmful interference to other fixed operations. Coordinating a repeater or frequency with other fixed operations demonstrates good engineering and amateur practice.

UK Repeaters

In the UK, the frequency allocations for repeaters are managed by the Emerging Technology Co-ordination Committee (ETCC) of the Radio Society of Great Britain and licensed by Ofcom, the industry regulator for communications in the UK. Each repeater has a NOV (Notice of Variation) licence issued to a particular amateur radio callsign (this person is normally known as the "repeater keeper") thus ensuring the licensing authority has a single point of contact for that particular repeater.

Each repeater in the UK is normally supported by a repeater group composed of local amateur radio enthusiasts who pay a nominal amount e.g. £10—15 a year each to support the maintenance of each repeater and to pay for site rents, electricity costs etc. Repeater groups do not receive any central funding from other organisations.

Such groups include the Central Scotland FM Group and the Scottish Borders Repeater Group.

Repeater Equipment

The most basic repeater consists of an FM receiver on one frequency and an FM transmitter on another frequency usually in the same radio band, connected together so that when the receiver picks up a signal, the transmitter is keyed and rebroadcasts whatever is heard.

2 Meter GE Mastr II repeater

In order to run the repeater a repeater controller is necessary. A repeater controller can be a hardware solution or even be implemented in software.

Repeaters typically have a timer to cut off retransmission of a signal that goes too long. Repeaters operated by groups with an emphasis on emergency communications often limit each transmission to 30 seconds, while others may allow three minutes or even longer. The timer restarts after a short pause following each transmission, and many systems feature a beep or chirp tone to signal that the timeout timer has reset.

Repeater Types

Conventional Repeaters

Conventional repeaters, also known as in-band or same-band repeaters, retransmit signals within the same frequency band, and they only repeat signals using a particular modulation scheme, predominately FM.

Standard repeaters require either the use of two antennas (one each for transmitter and receiver) or a duplexer to isolate the transmit and receive signals over a single antenna. The duplexer is a device which prevents the repeater's high-power transmitter (on the output frequency) from drowning out the users' signal on the repeater receiver (on the input frequency). A *diplexer* allows two transmitters on different frequencies to use one antenna, and is common in installations where one repeater on 2 m and a second on 440 MHz share one feedline up the tower and one antenna.

Most repeaters are remotely controlled through the use of audio tones on a control channel.

Cross-band Repeaters

A cross-band repeater (also sometimes called a replexer), is a repeater that retransmits a specific mode on a frequency in one band to a specific mode on a frequency in a different band. This technique allows for a smaller and less complex repeater system. Repeating signals across widely separated frequency bands allows for simple filters to be used to allow one antenna to be used for both transmit and receive at the same time. This avoids the use of complex duplexers to achieve the required rejection for same band repeating.

Some dual-band amateur transceivers are capable of cross-band repeat.

Amateur Television Repeaters

Amateur television (ATV) repeaters are used by amateur radio operators to transmit full motion video. The bands used by ATV repeaters vary by country, but in the US a typical configuration is as a cross-band system with an input on the 33 or 23 cm band and output on 421.25 MHz or, sometimes, 426.25 MHz (within the 70 cm band). These output frequencies happen to be the same as standard cable television channels 57 and 58, meaning that anyone with a cable-ready analog NTSC TV can tune them in without special equipment.

There are also digital amateur TV repeaters that retransmit digital video signals. Frequently DVB-S modulation is used for digital ATV, due to narrow bandwidth needs and high loss tolerances. These DATV repeaters are more prevalent in Europe currently, partially because of the availability of DVB-S equipment.

Satellite Repeaters

In addition, amateur radio satellites have been launched with the specific purpose of operating as space-borne amateur repeaters. The worldwide amateur satellite organization AMSAT designs and builds many of the amateur satellites, which are also known as OSCARs. Several satellites with amateur radio equipment on board have been designed and built by universities around the world. Also, several OSCARs have been built for experimentation. For example, NASA and AMSAT coordinated the release of SuitSat which was an attempt to make a low cost experimental satellite from a discarded Russian spacesuit outfitted with amateur radio equipment.

The repeaters on board a satellite may be of any type; the key distinction is that they are in orbit around the Earth, rather than terrestrial in nature. The three most common types of OSCARs are linear transponders, cross-band FM repeaters, and digipeaters (also referred to as pacsats).

Linear Transponders

Amateur transponder repeaters are most commonly used on amateur satellites. A specified band of frequencies, usually having a bandwidth of 20 to 800 kHz is repeated from one band to another. Transponders are not mode specific and typically no demodulation occurs. Any signal with a bandwidth narrower than the transponder's pass-band will be repeated; however, for technical reasons, use of modes other than SSB and CW are discouraged. Transponders may be inverting or non-inverting. An example of an inverting transponder would be a 70cm to 2m transponder which receives on the 432.000 MHz to 432.100 MHz frequencies and transmits on the 146.000 MHz to 146.100 MHz frequencies by inverting the frequency range within the band. In this example, a signal received at 432.001 MHz would be transmitted on 146.099 MHz. Voice signals using USB modulation on the input would result in a LSB modulation on the output, and vice versa.*"Phase 3D Satellite Primer"*.

Store-and-forward Systems

Another class of repeaters do not simultaneously retransmit a signal, on different frequency, as they receive it. Instead, they operate in a store-and-forward manner, by receiving and then re-transmitting on the same frequency after a short delay.

These systems may not be legally classified as "repeaters", depending on the definition set by a country's regulator. For example, in the US, the FCC defines a repeater as an "amateur station that simultaneously retransmits the transmission of another amateur station on a different channel or channels." (CFR 47 97.205(b)) Store-and-forward systems neither retransmit simultaneously, nor use a different channel. Thus, they must be operated under different rules than more conventional repeaters.

Simplex Repeaters

A type of system known as a *simplex repeater* uses a single transceiver and a short-duration voice recorder, which records whatever the receiver picks up for a set length of time (usually 30 seconds or less), then plays back the recording over the transmitter on the same frequency. A common name for them is a "parrot" repeater.

Digipeaters

Another form of repeater is used in amateur packet radio, a form of digital computer-to-computer communications, and are dubbed "digipeaters" (for *DIGItal rePEATERS)*. These repeaters are used for activities and modes such as packet radio, Automatic Packet Reporting System, and D-STAR's digital data mode.

SSTV Repeaters

An SSTV repeater is an amateur radio repeater station for relaying of slow-scan television signals. A typical SSTV repeater is equipped with a HF or VHF transceiver and a computer with a sound card, which serves as a demodulator/modulator of SSTV signals.

SSTV repeaters are used by amateur radio operators for exchanging pictures. If two stations can not copy each other, they can still communicate through a repeater.

To activate a repeater the station must send a tone of frequency 1,750 Hz. Then the repeater is activated and sends K in morse code. The station must start sending a picture in approximately 10 seconds. After reception the received image is transmitted on the repeater's operation frequency.

Repeaters should operate in common SSTV modes, but it depends on the software used (MMSSTV, JVComm32, MSCAN). Some repeater are not activated by audio tone, but instead by the SSTV vertical synchronization signal (VIS code).

Repeater Networks

Repeaters may be linked together in order to form what is known as a *linked repeater system* or *linked repeater network*. In such a system, when one repeater is keyed-up by receiving a signal, all the other repeaters in the network are also activated and will transmit the same signal. The connections between the repeaters are made via radio (usually on a different frequency from the published transmitting frequency) for maximum reliability. Some networks have a feature to allow the user being able to turn additional repeaters and links on or off on the network. This feature is typically done with DTMF tones to control the network infrastructure. Such a system allows coverage over a wide area, enabling communication between amateurs often hundreds of miles (several hundred km) apart. These systems are used for area or regional communications, for example in Skywarn nets, where storm spotters relay severe weather reports. All the user has to know is which channel to use in which area.

Voting Systems

In order to get better receive coverage over a wide area, a similar linked setup can also be done with what is known as a *voted receiver system*. In a voted receiver, there are several satellite receivers set up to receive on the same frequency (the one that the users transmit on). All of the satellite receivers are linked to a voting selector panel that switches from receiver to receiver based on the best quieting (strongest) signal, and the output of the selector will actually trigger the central repeater transmitter. A properly adjusted voting system can switch many times a second and can actually "assemble" a multi-syllable word using a different satellite receiver for each syllable. Such a system can be used to widen coverage to low power mobile radios or handheld radios that otherwise would not be able to key up the central location, but can receive the signal from the central location without an issue. Voting systems require no knowledge or effort on the part of the user - the system just seems to have better-than-average handheld coverage.

Internet Linking

Repeaters may also be connected to over the Internet using voice over IP (VoIP) techniques. VoIP links are a convenient way to connecting distant repeaters that would otherwise be unreachable by VHF/UHF radio propagation. Popular VoIP amateur radio network protocols include D-STAR, Echolink, IRLP, WIRES and eQSO.

Operating Terms

Timing Out is the situation where a person talks too long and the repeater timer shuts off the repeater transmitter.

Kerchunking is transmitting a momentary signal to check a repeater without identifying.

In many countries, such an act violates amateur radio regulations.

The term "Kerchunk" can also apply to the sound a large Amplitude Modulation Transmitter makes when the operator switches it off and on.

LID refers to a poor operator (radio methods) usually from improper training from other Amateurs or exposure to difference types of operation such as CB radio.

Absorption Wavemeter

An absorption wavemeter is a simple electronic instrument used to measure the frequency of radio waves. It is an older method of measuring frequency, widely used from the birth of radio in the early 20th century until the 1970s, when the development of inexpensive frequency counters, which have far greater accuracy, made it largely obsolete. A wavemeter consists of an adjustable resonant circuit calibrated in frequency, with a meter or other means to measure the voltage or current in the circuit. When adjusted to resonance with the unknown frequency, the resonant circuit absorbs energy, which is indicated by a dip on the meter. Then the frequency can be read from the dial.

Wavemeters are used for frequency measurements that do not require high accuracy, such as checking that a radio transmitter is operating within its correct frequency band, or checking for harmonics in the output. Many radio amateurs keep them as a simple way to check their output frequency. Similar devices can be made for detection of mobile phones. As an alternative, a dip meter can be used.

There are two categories of wavemeters: *transmission wavemeters*, which have an input and an output port and are inserted into the signal path, or *absorption wavemeters*, which are loosely coupled to the radio frequency source and absorb energy from it.

HF and VHF

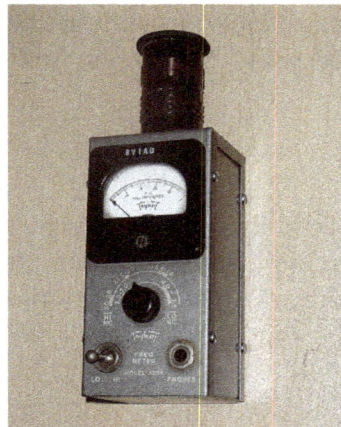

A Triplet 3256 wavemeter for use in the high frequency band.

The most simple form of the device is a variable capacitor with a coil wired across its terminals. Attached to one the terminals of the LC circuit is a diode, then between the end of the diode not wired to the LC circuit and the terminal of the LC circuit not bearing the diode is wired a ceramic decoupling capacitor. Finally a galvanometer is wired to the terminals of the decoupling capacitor.

The device will be sensitive to strong sources of radiowaves at the frequency at which the LC circuit is resonant.

This is given by $f = \dfrac{1}{2\pi\sqrt{LC}}$

When the device is exposed to an RF field which is at the resonant frequency a DC voltage will appear on the terminals on the left hand side. The coil is often outside the case of the unit so it can be brought close to the object being probed.

UHF and SHF

At the higher frequencies it is not possible to use lumped components for the tuned circuit. Instead methods such as stripline or resonant cavities are used. One design for ultra high frequencies (UHF) and super high frequencies (SHF) is a resonant λ/4 (quarter wave) rod which can vary in length. Another design for X-band (10 GHz) is a resonant cavity which can be changed in length.

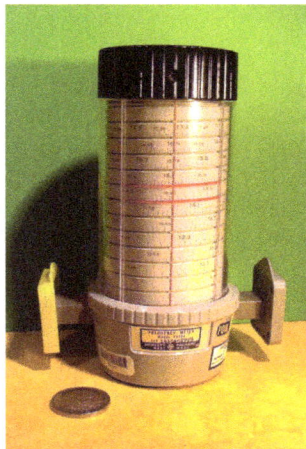

Resonant cavity wavemeter for measuring microwave frequencies in the K_u band

As an alternative for UHF, Lecher transmission lines can be used. It is possible to measure roughly the frequency of a transmitter using Lecher lines.

Lecher Lines

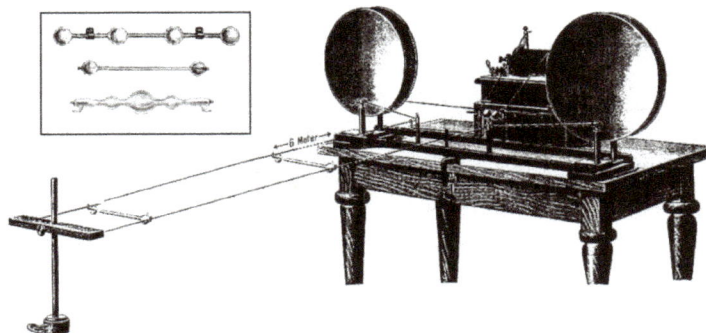

Early 1902 Lecher line identical to Ernst Lecher's original 1888 apparatus. Radio waves generated by the Hertzian spark-gap oscillator at right travel down the parallel wires. The wires are short-circuited together at the left end, reflecting the waves back up the wires toward the oscillator, creating a standing wave of voltage along the line. The voltage goes to zero at nodes located at multiples of a half-wavelength from the end. The nodes were found by sliding a Geissler tube, a small glow discharge tube like a neon light, up and down the line. The high voltage on the line makes the tube glow. When the tube reaches a node, the voltage goes to zero and the tube goes out. The measured distance between two successive nodes is multiplied by two to get the wavelength λ of the radio waves. The line is shown truncated in the drawing; the length of the line was actually 6 meters (18 feet). The radio waves produced by the oscillator were in the UHF range, with a wavelength of several meters. The inset shows types of Geissler tube used with Lecher lines.

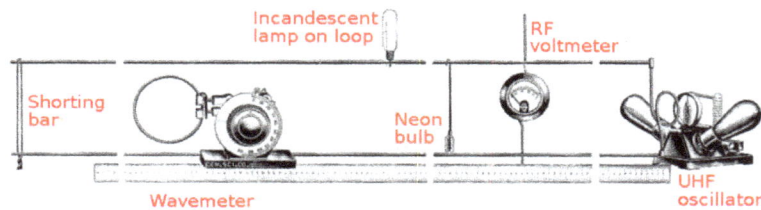

Lecher-line educational kit sold by Central Scientific Co. in the 1930s for teaching radio theory in college. It contains everything necessary, including an absorption wavemeter for independently measuring frequency.

In electronics, a Lecher line or Lecher wires is a pair of parallel wires or rods that were used to measure the wavelength of radio waves, mainly at UHF and microwave frequencies. They form a short length of balanced transmission line (a resonant stub). When attached to a source of radio-frequency power such as a radio transmitter, the radio waves form standing waves along their length. By sliding a conductive bar that bridges the two wires along their length, the length of the waves can be physically measured. Austrian physicist Ernst Lecher, improving on techniques used by Oliver Lodge and Heinrich Hertz, developed this method of measuring wavelength around 1888. Lecher lines were used as frequency measuring devices until frequency counters became available after World War 2. They were also used as components, often called "resonant stubs", in UHF and microwave radio equipment such as transmitters, radar sets, and television sets, serving as tank circuits, filters, and impedance-matching devices. They are used at frequencies between HF/VHF, where lumped components are used, and UHF/SHF, where resonant cavities are more practical.

Wavelength Measurement

A Lecher line is a pair of parallel uninsulated wires or rods held a precise distance apart. The separation is not critical but should be a small fraction of the wavelength; it ranges from less than a centimeter to over 10 cm. The length of the wires depends on the wavelength involved; lines used for measurement are generally several wavelengths long. The uniform spacing of the wires makes them a transmission line, conducting radio waves at a constant speed very close to the speed of light. One end of the rods is connected to the source of RF power, such as the output of a radio transmitter. At the other end the rods are connected together with a conductive bar between them. This short circuiting termination reflects the waves. The waves reflected from the short-circuited end interfere with the outgoing waves, creating a sinusoidal standing wave of voltage and current on the line. The voltage goes to zero at nodes located at multiples of half a wavelength from the end, with maxima called antinodes located midway between the nodes.

Therefore, the wavelength λ can be determined by finding the location of two successive nodes (or antinodes) and measuring the distance between them, and multiplying by two. The frequency f of the waves can be calculated from the wavelength and the speed of the waves, which is the speed of light c:

$$f = \frac{c}{\lambda}$$

The nodes are much sharper than the antinodes, because the change of voltage with distance along the line is maximum at the nodes, so they are used.

Finding the Nodes

Two methods are employed to find the nodes. One is to use some type of voltage indicator, such as an RF voltmeter or light bulb, attached to a pair of contacts that slide up and down the wires. When the bulb reaches a node, the voltage between the wires goes to zero, so the bulb goes out. If the indicator has too low an impedance it will disturb the standing wave on the line, so a high impedance indicator must be used; a regular incandescent bulb has too low a resistance. Lecher and early researchers used long thin Geissler tubes, laying the glass tube directly across the line. The high voltage of early transmitters excited a glow discharge in the gas. In modern times small neon bulbs are often used. One problem with using glow discharge bulbs is their high striking voltage makes it difficult to localize the exact voltage minimum. In precision wavemeters an RF voltmeter is used.

The other method used to find the nodes is to slide the terminating shorting bar up and down the line, and measure the current flowing into the line with an RF ammeter in the feeder line. The current on the Lecher line, like the voltage, forms a standing wave with nodes (points of minimum current) every half wavelength. So the line presents an impedance to the applied power which varies with its length; when a current node is located at the entrance to the line, the current drawn from the source, measured by the ammeter, will be minimum. The shorting bar is slid down the line and the position of two successive current minima is noted, the distance between them is half a wavelength.

With care, Lecher lines can measure frequency to an accuracy of 0.1%.

Construction

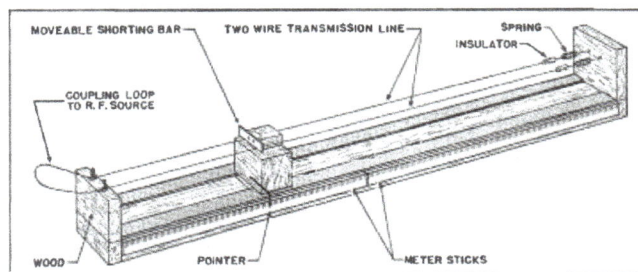

Lecher line wavemeter, from "DIY" article in 1946 radio magazine

A major attraction of Lecher lines was they were a way to measure frequency without complicated electronics, and could be improvised from simple materials found in a typical shop. Lecher line wavemeters are usually built on a frame which holds the conductors rigid and horizontal, with a track that the shorting bar or indicator rides on, and a built-in measuring scale so the distance between nodes can be read out. The frame must be made of a nonconductive material like wood, because any conducting objects near the line can disturb the standing wave pattern. The RF current is usually coupled into the line through a single turn loop of wire at one end, which can be held near a transmitter's tank coil.

A simpler design is a "U"-shaped metal bar, marked with graduations, with a sliding shorting bar. In operation, the U end acts as a coupling link and is held near the transmitter's tank coil, and the shorting bar is slid out along the arms until the transmitter's plate current dips, indicating the first node has been reached. Then the distance from the end of the link to the shorting bar is a half-wavelength. The shorting bar should always be slid *out*, away from the link end, not *in*, to avoid converging on a higher order node by mistake.

In many ways Lecher lines are an electrical version of the Kundt's tube experiment which is used to measure the wavelength of sound waves.

Measuring the Speed of Light

If the frequency f of the radio waves is independently known, the wavelength λ measured on a Lecher line can be used to calculate the speed of the waves, c, which is approximately equal to the speed of light:

$$c = \lambda f$$

In 1891, French physicist Prosper-René Blondlot made the first measurement of the speed of radio waves, using this method. He used 13 different frequencies between 10 and 30 MHz and obtained an average value of 297,600 km/s, which is within 1% of the current value for the speed of light. Other researchers repeated the experiment with greater accuracy. This was an important confirmation of James Clerk Maxwell's theory that light was an electromagnetic wave like radio waves.

Other Applications

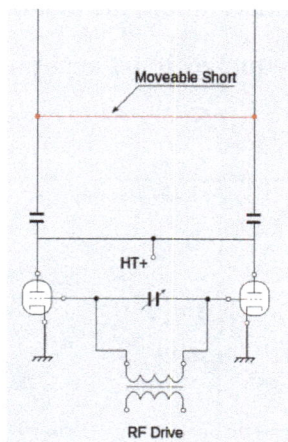

Lecher line as a tank circuit in an RF amplifier. Not shown in this simplified diagram are the chokes that feed the tube anodes from the HT source. Without them the two anodes are shorted together.

Short lengths of Lecher line are often used as high Q resonant circuits, termed *resonant stubs*. For example, a quarter wavelength ($\lambda/4$) shorted Lecher line acts like a parallel resonant circuit, appearing as a high impedance at its resonant frequency and low impedance at other frequencies. They are used because at UHF frequencies the value of inductors and capacitors needed for 'lumped component' tuned circuits becomes extremely low, making them difficult to fabricate and sensitive to parasitic capacitance and inductance. One difference between them is that transmission line stubs like Lecher lines also resonate at odd-number multiples of their fundamental resonant frequency, while lumped LC circuits just have one resonant frequency.

Power Amplifier Tank Circuits

Lecher line circuits can be used for the tank circuits of UHF power amplifiers. For instance, the twin tetrode (QQV03-20) 432 MHz amplifier described by G.R Jessop uses a Lecher line anode tank.

Television Tuners

Quarter-wave Lecher lines are used for the tuned circuits in the RF amplifier and local oscillator portions of modern television sets. The tuning necessary to select different stations is done by varactor diodes across the Lecher line.

Characteristic Impedance of Lecher Line

The separation between the Lecher bars does not affect the position of the standing waves on the line, but it does determine the characteristic impedance, which can be important for matching the line to the source of the radio frequency energy for efficient power transfer. For two parallel cylindrical conductors of diameter d and spacing D,

$$Z_0 = 276 \log\left(\frac{D}{d} + \sqrt{\left(\frac{D}{d}\right)^2 - 1}\right) = \frac{120}{\sqrt{\epsilon_r}} \cosh^{-1}\left(\frac{D}{d}\right)$$

For parallel wires the formula for capacitance is

- l, length

- C, capacitance per meter

$$C = \frac{\pi \epsilon_0 \epsilon_r}{\ln\left(\frac{2D}{d}\right)}$$

Hence as

$$Z_0^2 = \frac{L}{C}$$

$$c = \frac{1}{\sqrt{\dfrac{L}{C} \cdot C^2}}$$

$$= \frac{1}{Z_0 \cdot \left(\pi \epsilon_0 \epsilon_r \right) \cdot \ln\left(\dfrac{2D}{d} \right)}$$

Commercially available 300 and 450 ohm twin lead balanced ribbon feeder can be used as a fixed length Lecher line (resonant stub).

References

- Lee, Thomas H. (2004). The Design of CMOS Radio-Frequency Integrated Circuits, 2nd Ed. UK: Cambridge University Press. pp. 1–8. ISBN 0521835399

- Armstrong, Edwin H. (February 1921). "A new system of radio frequency amplification". Proceedings of the Inst. of Radio Engineers. New York: Institute of Radio Engineers. 9 (1): 3–11. Retrieved December 23, 2015

- Aitken, Hugh G.J. (2014). The Continuous Wave: Technology and American Radio, 1900-1932. Princeton Univ. Press. p. 190. ISBN 1400854601

- Hogan, John V. L. (April 1921). "The Heterodyne Receiver". The Electric Journal. Pittsburgh, USA: The Electric Journal. 18 (4): 116–119. Retrieved January 28, 2016

- Maver, William Jr. (August 1904). "Wireless Telegraphy To-Day". American Monthly Review of Reviews. New York: The Review of Reviews Co. 30 (2): 192. Retrieved January 2, 2016

- Rockman, Howard B. (2004). Intellectual Property Law for Engineers and Scientists. John Wiley and Sons. pp. 196–199. ISBN 0471697397

- Crookes, William (February 1, 1892). "Some Possibilities of Electricity". The Fortnightly Review. London: Chapman and Hall. 51: 174–176. Retrieved August 19, 2015

- Wurtzler, Steve J. (2007). Electric Sounds: Technological Change and the Rise of Corporate Mass Media. Columbia Univ. Press. pp. 147–148. ISBN 023151008X

- Grimes, David (May 1924). "The Story of Reflex and Radio Frequency" (PDF). Radio in the Home. 2 (12): 9–10,. Retrieved January 24, 2016

- Armstrong, Edwin H. (April 1921). "The Regenerative Circuit". The Electrical Journal. Pittsburgh, PA: Westinghouse Co. 18 (4): 153–154. Retrieved January 11, 2016

- Petruzellis, Thomas (2007). 22 Radio and Receiver Projects for the Evil Genius. US: McGraw-Hill Professional. pp. 40, 44. ISBN 978-0-07-148929-4

- Kuhn, Kenneth A. (Jan 6, 2008). "Introduction" (PDF). Crystal Radio Engineering. Prof. Kenneth Kuhn website, Univ. of Alabama. Retrieved 2009-12-07

- Williams, Lyle R. (2006). The New Radio Receiver Building Handbook. The Alternative Electronics Press. pp. 20–23. ISBN 978-1-84728-526-3

Understanding Radio Antenna and its Types

Antennas are devices used to convert electrical energy into radio waves. Antennas are mainly used in radio receivers and radio transmitters. Some of the types of radio antennas are dipole antenna, adcock antenna, conformal antenna, dielectric resonator antenna and cage aerial. Radio antenna is best understood in confluence with the major topics listed in the following chapter.

Antenna (Radio)

In radio and electronics, an antenna (plural antennae or antennas), or aerial, is an electrical device which converts electric power into radio waves, and vice versa. It is usually used with a radio transmitter or radio receiver. In transmission, a radio transmitter supplies an electric current to the antenna's terminals, and the antenna radiates the energy from the current as electromagnetic waves (radio waves). In reception, an antenna intercepts some of the power of an electromagnetic wave in order to produce an electric current at its terminals, that is applied to a receiver to be amplified.

Antennas are essential components of all equipment that uses radio. They are used in systems such as radio broadcasting, broadcast television, two-way radio, communications receivers, radar, cell phones, and satellite communications, as well as other devices such as garage door openers, wireless microphones, Bluetooth-enabled devices, wireless computer networks, baby monitors, and RFID tags on merchandise.

Typically an antenna consists of an arrangement of metallic conductors (elements), electrically connected (often through a transmission line) to the receiver or transmitter. An oscillating current of electrons forced through the antenna by a transmitter will create an oscillating magnetic field around the antenna elements, while the charge of the electrons also creates an oscillating electric field along the elements. These time-varying fields radiate away from the antenna into space as a moving transverse electromagnetic field wave. Conversely, during reception, the oscillating electric and magnetic fields of an incoming radio wave exert force on the electrons in the antenna elements, causing them to move back and forth, creating oscillating currents in the antenna.

Antennas can be designed to transmit and receive radio waves in all horizontal directions equally (omnidirectional antennas), or preferentially in a particular direction (directional or high gain antennas). In the latter case, an antenna may also include additional elements or surfaces with no electrical connection to the transmitter or receiver, such as parasitic elements, parabolic reflectors or horns, which serve to direct the radio waves into a beam or other desired radiation pattern.

The first antennas were built in 1888 by German physicist Heinrich Hertz in his pioneering experiments to prove the existence of electromagnetic waves predicted by the theory of James Clerk

Maxwell. Hertz placed dipole antennas at the focal point of parabolic reflectors for both transmitting and receiving.

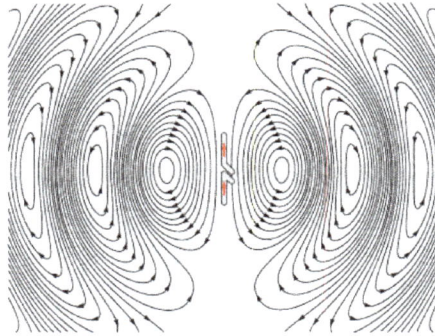

This figure is a half-wave dipole antenna transmitting radio waves, showing the electric field lines. The antenna in the center is two vertical metal rods, with an alternating current applied at its center from a radio transmitter *(not shown)*. The voltage charges the two sides of the antenna alternately positive *(+)* and negative *(−)*. Loops of electric field *(black lines)* leave the antenna and travel away at the speed of light; these are the radio waves.

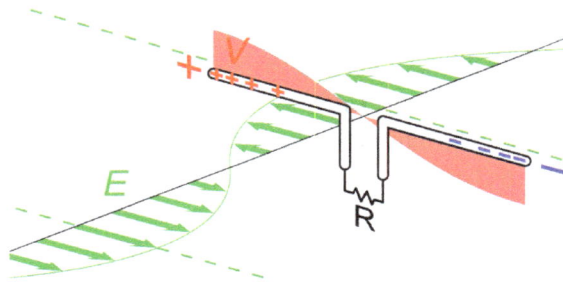

This diagram of a half-wave dipole antenna receiving energy from a radio wave. The antenna consists of two metal rods connected to a receiver *R*. The electric field *(E, green arrows)* of the incoming wave pushes the electrons in the rods back and forth, charging the ends alternately positive *(+)* and negative *(−)*. Since the length of the antenna is one half the wavelength of the wave, the oscillating field induces standing waves of voltage *(V, represented by red band)* and current in the rods. The oscillating currents *(black arrows)* flow down the transmission line and through the receiver (represented by the resistance *R*).

Terminology

Electronic symbol for an antenna

The words *antenna* (plural: *antennas* in US English, although both "antennas" and "antennae" are used in International English) and *aerial* are used interchangeably. Occasionally the term "aerial" is used to mean a wire antenna. However, note the important international technical journal, the

IEEE Transactions on Antennas and Propagation. In the United Kingdom and other areas where British English is used, the term aerial is sometimes used although 'antenna' has been universal in professional use for many years.

The origin of the word *antenna* relative to wireless apparatus is attributed to Italian radio pioneer Guglielmo Marconi. In the summer of 1895, Marconi began testing his wireless system outdoors on his father's estate near Bologna and soon began to experiment with long wire "aerials". Marconi discovered that by raising the "aerial" wire above the ground and connecting the other side of his transmitter to ground, the transmission range was increased. Soon he was able to transmit signals over a hill, a distance of approximately 2.4 kilometres (1.5 mi). In Italian a tent pole is known as *l'antenna centrale,* and the pole with the wire was simply called *l'antenna.* Until then wireless radiating transmitting and receiving elements were known simply as aerials or terminals. Because of his prominence, Marconi's use of the word *antenna* spread among wireless researchers, and later to the general public.

In common usage, the word *antenna* may refer broadly to an entire assembly including support structure, enclosure (if any), etc. in addition to the actual functional components. Especially at microwave frequencies, a receiving antenna may include not only the actual electrical antenna but an integrated preamplifier or mixer.

An antenna, in converting radio waves to electrical signals or vice versa, is a form of transducer.

Overview

Antennas of the Atacama Large Millimeter submillimeter Array.

Antennas are required by any radio receiver or transmitter to couple its electrical connection to the electromagnetic field. Radio waves are electromagnetic waves which carry signals through the air (or through space) at the speed of light with almost no transmission loss. Radio transmitters and receivers are used to convey signals (information) in systems including broadcast (audio) radio, television, mobile telephones, Wi-Fi (WLAN) data networks, trunk lines and point-to-point communications links (telephone, data networks), satellite links, many remote controlled devices such as garage door openers, and wireless remote sensors, among many others. Radio waves are also used directly for measurements in technologies including radar, GPS, and radio astronomy. In each and every case, the transmitters and receivers involved require antennas, although these are sometimes hidden (such as the antenna inside an AM radio or inside a laptop computer equipped with Wi-Fi).

Whip antenna on car, common example of an omnidirectional antenna

According to their applications and technology available, antennas generally fall in one of two categories:

1. Omnidirectional or only weakly directional antennas which receive or radiate more or less in all directions. These are employed when the relative position of the other station is unknown or arbitrary. They are also used at lower frequencies where a directional antenna would be too large, or simply to cut costs in applications where a directional antenna isn't required.

2. Directional or *beam* antennas which are intended to preferentially radiate or receive in a particular direction or directional pattern.

In common usage "omnidirectional" usually refers to all horizontal directions, typically with reduced performance in the direction of the sky or the ground (a truly isotropic radiator is not even possible). A "directional" antenna usually is intended to maximize its coupling to the electromagnetic field in the direction of the other station, or sometimes to cover a particular sector such as a 120° horizontal fan pattern in the case of a panel antenna at a cell site.

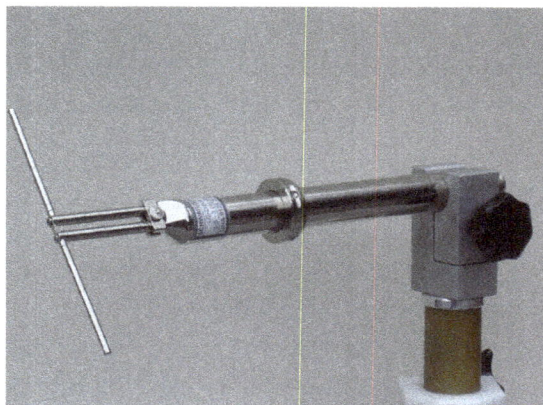

Half-wave dipole antenna

One example of omnidirectional antennas is the very common *vertical antenna* or whip antenna consisting of a metal rod (often, but not always, a quarter of a wavelength long). A dipole antenna is similar but consists of two such conductors extending in opposite directions, with a total length that is often, but not always, a half of a wavelength long. Dipoles are typically oriented horizontally in which case they are weakly directional: signals are reasonably well radiated toward or received from all directions with the exception of the direction along the conductor itself; this region is called the antenna blind cone or null.

Both the vertical and dipole antennas are simple in construction and relatively inexpensive. The dipole antenna, which is the basis for most antenna designs, is a balanced component, with equal but opposite voltages and currents applied at its two terminals through a balanced transmission line (or to a coaxial transmission line through a so-called balun). The vertical antenna, on the other hand, is a *monopole* antenna. It is typically connected to the inner conductor of a coaxial transmission line (or a matching network); the shield of the transmission line is connected to ground. In this way, the ground (or any large conductive surface) plays the role of the second conductor of a dipole, thereby forming a complete circuit. Since monopole antennas rely on a conductive ground, a so-called grounding structure may be employed to provide a better ground contact to the earth or which itself acts as a ground plane to perform that function regardless of (or in absence of) an actual contact with the earth.

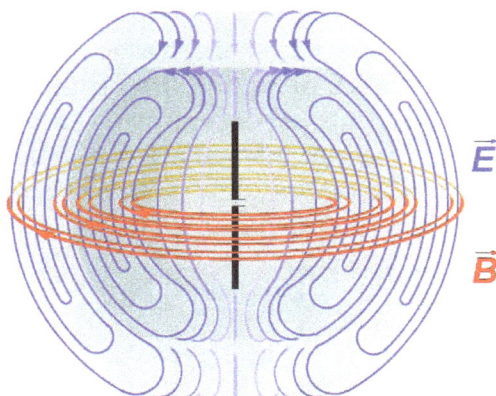

Diagram of the electric fields *(blue)* and magnetic fields *(red)* radiated by a dipole antenna *(black rods)* during transmission.

Antennas more complex than the dipole or vertical designs are usually intended to increase the directivity and consequently the gain of the antenna. This can be accomplished in many different ways leading to a plethora of antenna designs. The vast majority of designs are fed with a balanced line (unlike a monopole antenna) and are based on the dipole antenna with additional components (or *elements*) which increase its directionality. Antenna "gain" in this instance describes the concentration of radiated power into a particular solid angle of space, as opposed to the spherically uniform radiation of the ideal radiator. The increased power in the desired direction is at the expense of that in the undesired directions. Power is conserved, and there is no net power increase over that delivered from the power source (the transmitter.)

For instance, a phased array consists of two or more simple antennas which are connected together through an electrical network. This often involves a number of parallel dipole antennas with a certain spacing. Depending on the relative phase introduced by the network, the same combination of dipole antennas can operate as a "broadside array" (directional normal to a line connecting the elements) or as an "end-fire array" (directional along the line connecting the elements). Antenna arrays may employ any basic (omnidirectional or weakly directional) antenna type, such as dipole, loop or slot antennas. These elements are often identical.

However a log-periodic dipole array consists of a number of dipole elements of *different* lengths in order to obtain a somewhat directional antenna having an extremely wide bandwidth: these are frequently used for television reception in fringe areas. The dipole antennas composing it are all considered "active elements" since they are all electrically connected together (and to the transmission line). On the other hand, a superficially similar dipole array, the Yagi-Uda Antenna (or simply "Yagi"), has only one dipole element with an electrical connection; the other so-called parasitic elements interact with the electromagnetic field in order to realize a fairly directional antenna but one which is limited to a rather narrow bandwidth. The Yagi antenna has similar looking parasitic dipole elements but which act differently due to their somewhat different lengths. There may be a number of so-called "directors" in front of the active element in the direction of propagation, and usually a single (but possibly more) "reflector" on the opposite side of the active element.

Greater directionality can be obtained using beam-forming techniques such as a parabolic reflector or a horn. Since high directivity in an antenna depends on it being large compared to the wavelength, narrow beams of this type are more easily achieved at UHF and microwave frequencies.

At low frequencies (such as AM broadcast), arrays of vertical towers are used to achieve directionality and they will occupy large areas of land. For reception, a long Beverage antenna can have significant directivity. For non directional portable use, a short vertical antenna or small loop antenna works well, with the main design challenge being that of impedance matching. With a vertical antenna a *loading coil* at the base of the antenna may be employed to cancel the reactive component of impedance; small loop antennas are tuned with parallel capacitors for this purpose.

Cell phone base station antennas

An antenna lead-in is the transmission line (or *feed line*) which connects the antenna to a transmitter or receiver. The *antenna feed* may refer to all components connecting the antenna to the transmitter or receiver, such as an impedance matching network in addition to the transmission line. In a so-called aperture antenna, such as a horn or parabolic dish, the "feed" may also refer to a basic antenna inside the entire system (normally at the focus of the parabolic dish or at the throat of a horn) which could be considered the one active element in that antenna system. A microwave antenna may also be fed directly from a waveguide in place of a (conductive) transmission line.

An antenna counterpoise or ground plane is a structure of conductive material which improves or substitutes for the ground. It may be connected to or insulated from the natural ground. In a monopole antenna, this aids in the function of the natural ground, particularly where variations (or limitations) of the characteristics of the natural ground interfere with its proper function. Such a structure is normally connected to the return connection of an unbalanced transmission line such as the shield of a coaxial cable.

An electromagnetic wave *refractor* in some aperture antennas is a component which due to its shape and position functions to selectively delay or advance portions of the electromagnetic wavefront passing through it. The refractor alters the spatial characteristics of the wave on one side relative to the other side. It can, for instance, bring the wave to a focus or alter the wave front in other ways, generally in order to maximize the directivity of the antenna system. This is the radio equivalent of an optical lens.

An antenna coupling network is a passive network (generally a combination of inductive and capacitive circuit elements) used for impedance matching in between the antenna and the transmitter or receiver. This may be used to improve the standing wave ratio in order to minimize losses in the transmission line and to present the transmitter or receiver with a standard resistive impedance that it expects to see for optimum operation.

Reciprocity

It is a fundamental property of antennas that the electrical characteristics of an antenna described in the next section, such as gain, radiation pattern, impedance, bandwidth, resonant frequency and polarization, are the same whether the antenna is transmitting or receiving. For example, the *"receiving pattern"* (sensitivity as a function of direction) of an antenna when used for reception is identical to the radiation pattern of the antenna when it is *driven* and functions as a radiator. This is a consequence of the reciprocity theorem of electromagnetics. Therefore, in discussions of antenna properties no distinction is usually made between receiving and transmitting terminology, and the antenna can be viewed as either transmitting or receiving, whichever is more convenient.

A necessary condition for the aforementioned reciprocity property is that the materials in the antenna and transmission medium are linear and reciprocal. *Reciprocal* (or *bilateral*) means that the material has the same response to an electric current or magnetic field in one direction, as it has to the field or current in the opposite direction. Most materials used in antennas meet these conditions, but some microwave antennas use high-tech components such as isolators and circulators, made of nonreciprocal materials such as ferrite. These can be used to give the antenna a different behavior on receiving than it has on transmitting, which can be useful in applications like radar.

Characteristics

Antennas are characterized by a number of performance measures which a user would be concerned with in selecting or designing an antenna for a particular application. Chief among these relate to the directional characteristics (as depicted in the antenna's *radiation pattern*) and the resulting *gain*. Even in omnidirectional (or weakly directional) antennas, the gain can often be increased by concentrating more of its power in the horizontal directions, sacrificing power radiated toward the sky and ground. The antenna's power gain (or simply "gain") also takes into account the antenna's efficiency, and is often the primary figure of merit.

Resonant antennas are expected to be used around a particular *resonant frequency*; an antenna must therefore be built or ordered to match the frequency range of the intended application. A particular antenna design will present a particular feedpoint impedance. While this may affect the choice of an antenna, an antenna's impedance can also be adapted to the desired impedance level of a system using a matching network while maintaining the other characteristics (except for a possible loss of efficiency).

Although these parameters can be measured in principle, such measurements are difficult and require very specialized equipment. Beyond tuning a transmitting antenna using an SWR meter, the typical user will depend on theoretical predictions based on the antenna design or on claims of a vendor.

An antenna transmits and receives radio waves with a particular polarization which can be reoriented by tilting the axis of the antenna in many (but not all) cases. The physical size of an antenna is often a practical issue, particularly at lower frequencies (longer wavelengths). Highly directional antennas need to be significantly larger than the wavelength. Resonant antennas usually use a linear conductor (or *element*), or pair of such elements, each of which is about a quarter of the wavelength in length (an odd multiple of quarter wavelengths will also be resonant). Antennas that are required to be small compared to the wavelength sacrifice efficiency and cannot be very directional. Fortunately at higher frequencies (UHF, microwaves) trading off performance to obtain a smaller physical size is usually not required.

Resonant Antennas

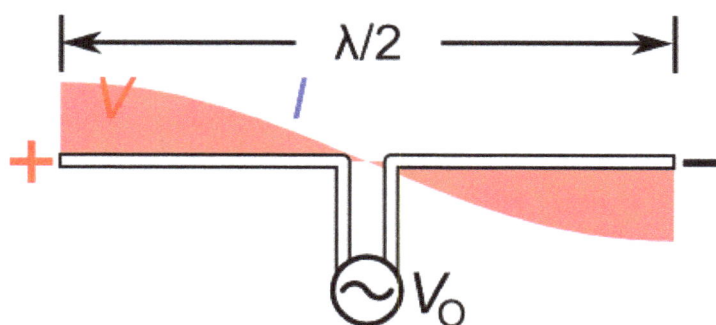

Standing waves on a half wave dipole driven at its resonant frequency. The waves are shown graphically by bars of color (red for voltage, *V* and blue for current, *I*) whose width is proportional to the amplitude of the quantity at that point on the antenna.

The majority of antenna designs are based on the *resonance* principle. This relies on the behaviour of moving electrons, which reflect off surfaces where the dielectric constant changes, in a fashion similar to the way light reflects when optical properties change. In these designs, the reflective surface is created by the end of a conductor, normally a thin metal wire or rod, which in the simplest case has a *feed point* at one end where it is connected to a transmission line. The conductor, or *ele-*

ment, is aligned with the electrical field of the desired signal, normally meaning it is perpendicular to the line from the antenna to the source (or receiver in the case of a broadcast antenna).

The radio signal's electrical component induces a voltage in the conductor. This causes an electrical current to begin flowing in the direction of the signal's instantaneous field. When the resulting current reaches the end of the conductor, it reflects, which is equivalent to a 180 degree change in phase. If the conductor is $\frac{1}{4}$ of a wavelength long, current from the feed point will undergo 90 degree phase change by the time it reaches the end of the conductor, reflect through 180 degrees, and then another 90 degrees as it travels back. That means it has undergone a total 360 degree phase change, returning it to the original signal. The current in the element thus adds to the current being created from the source at that instant. This process creates a standing wave in the conductor, with the maximum current at the feed.

The ordinary half-wave dipole is probably the most widely used antenna design. This consists of two $\frac{1}{4}$-wavelength elements arranged end-to-end, and lying along essentially the same axis (or *collinear*), each feeding one side of a two-conductor transmission wire. The physical arrangement of the two elements places them 180 degrees out of phase, which means that at any given instant one of the elements is driving current into the transmission line while the other is pulling it out. The monopole antenna is essentially one half of the half-wave dipole, a single $\frac{1}{4}$-wavelength element with the other side connected to ground or an equivalent ground plane (or *counterpoise*). Monopoles, which are one-half the size of a dipole, are common for long-wavelength radio signals where a dipole would be impractically large. Another common design is the folded dipole, which is essentially two dipoles placed side-by-side and connected at their ends to make a single one-wavelength antenna.

The standing wave forms with this desired pattern at the design frequency, f_o, and antennas are normally designed to be this size. However, feeding that element with $3f_o$ (whose wavelength is $\frac{1}{3}$ that of f_o) will also lead to a standing wave pattern. Thus, an antenna element is *also* resonant when its length is $\frac{3}{4}$ of a wavelength. This is true for all odd multiples of $\frac{1}{4}$ wavelength. This allows some flexibility of design in terms of antenna lengths and feed points. Antennas used in such a fashion are known to be *harmonically operated*.

Current and Voltage Distribution

The quarter-wave elements imitate a series-resonant electrical element due to the standing wave present along the conductor. At the resonant frequency, the standing wave has a current peak and voltage node (minimum) at the feed. In electrical terms, this means the element has minimum reactance, generating the maximum current for minimum voltage. This is the ideal situation, because it produces the maximum output for the minimum input, producing the highest possible efficiency. Contrary to an ideal (lossless) series-resonant circuit, a finite resistance remains (corresponding to the relatively small voltage at the feed-point) due to the antenna's radiation resistance as well as any actual electrical losses.

Recall that a current will reflect when there are changes in the electrical properties of the material. In order to efficiently send the signal into the transmission line, it is important that the transmission line has the same impedance as the elements, otherwise some of the signal will be reflected back into the antenna. This leads to the concept of impedance matching, the design of the overall system of antenna and transmission line so the impedance is as close as possible, thereby reducing these losses. Impedance matching between antennas and transmission lines is commonly handled through the use of a balun, although other solutions are also used in certain roles. An important measure of this basic concept is the standing wave ratio, which measures the magnitude of the reflected signal.

Consider a half-wave dipole designed to work with signals 1 m wavelength, meaning the antenna would be approximately 50 cm across. If the element has a length-to-diameter ratio of 1000, it will have an inherent resistance of about 63 ohms. Using the appropriate transmission wire or balun, we match that resistance to ensure minimum signal loss. Feeding that antenna with a current of 1 ampere will require 63 volts of RF, and the antenna will radiate 63 watts (ignoring losses) of radio frequency power. Now consider the case when the antenna is fed a signal with a wavelength of 1.25 m; in this case the reflected current would arrive at the feed out-of-phase with the signal, causing the net current to drop while the voltage remains the same. Electrically this appears to be a very high impedance. The antenna and transmission line no longer have the same impedance, and the signal will be reflected back into the antenna, reducing output. This could be addressed by changing the matching system between the antenna and transmission line, but that solution only works well at the new design frequency.

The end result is that the resonant antenna will efficiently feed a signal into the transmission line only when the source signal's frequency is close to that of the design frequency of the antenna, or one of the resonant multiples. This makes resonant antenna designs inherently narrowband, and they are most commonly used with a single target signal. They are particularly common on radar systems, where the same antenna is used for both broadcast and reception, or for radio and television *broadcasts*, where the antenna is working with a single frequency. They are less commonly used for reception where multiple channels are present, in which case additional modifications are used to increase the bandwidth, or entirely different antenna designs are used.

Electrically Short Antennas

It is possible to use simple impedance matching concepts to allow the use of monopole or dipole antennas substantially shorter than the ¼ or ½ wavelength, respectively, at which they are resonant. As these antennas are made shorter (for a given frequency) their impedance becomes dominated by a series capacitive (negative) reactance; by adding a series inductance with the opposite (positive) reactance – a so-called loading coil – the antenna's reactance may be cancelled leaving only a pure resistance. Sometimes the resulting (lower) electrical resonant frequency of such a system (antenna plus matching network) is described using the concept of *electrical length*, so an antenna used at a lower frequency than its resonant frequency is called an *electrically short antenna*.

For example, at 30 MHz (10 m wavelength) a true resonant ¼ wavelength monopole would be almost 2.5 meters long, and using an antenna only 1.5 meters tall would require the addition of a loading coil. Then it may be said that the coil has lengthened the antenna to achieve an electrical length of 2.5 meters. However, the resulting resistive impedance achieved will be quite a bit lower than that of a true ¼ wave (resonant) monopole, often requiring further impedance matching (a transformer) to the desired transmission line. For ever shorter antennas (requiring greater "electrical lengthening") the radiation resistance plummets (approximately according to the square of the antenna length), so that the mismatch due to a net reactance away from the electrical resonance worsens. Or one could as well say that the equivalent resonant circuit of the antenna system has a higher Q factor and thus a reduced bandwidth, which can even become inadequate for the transmitted signal's spectrum. Resistive losses due to the loading coil, relative to the decreased radiation resistance, entail a reduced electrical efficiency, which can be of great concern for a transmitting antenna, but bandwidth is the major factor that sets the size of antennas at 1 MHz and lower frequencies.

Arrays and Reflectors

Rooftop television Yagi-Uda antennas like these are widely used at VHF and UHF frequencies.

The amount of signal received from a distant transmission source is essentially geometric in nature due to the inverse-square law, and this leads to the concept of *effective area*. This measures the performance of an antenna by comparing the amount of power it generates to the amount of power in the original signal, measured in terms of the signal's power density in Watts per square metre. A half-wave dipole has an effective area of 0.13 ^{2}. If more performance is needed, one cannot simply make the antenna larger. Although this would intercept more energy from the signal, due to the considerations above, it would decrease the output significantly due to it moving away from the resonant length. In roles where higher performance is needed, designers often use multiple elements combined together.

Returning to the basic concept of current flows in a conductor, consider what happens if a half-wave dipole is not connected to a feed point, but instead shorted out. Electrically this forms a single $\frac{1}{2}$-wavelength element. But the overall current pattern is the same; the current will be zero at the two ends, and reach a maximum in the center. Thus signals near the design frequency will continue to create a standing wave pattern. Any varying electrical current, like the standing wave in the element, will radiate a signal. In this case, aside from resistive losses in the element, the rebroadcast signal will be significantly similar to the original signal in both magnitude and shape. If this element is placed so its signal reaches the main dipole in-phase, it will reinforce the original signal, and increase the current in the dipole. Elements used in this way are known as *passive elements*.

A Yagi-Uda array uses passive elements to greatly increase gain. It is built along a support boom that is pointed toward the signal, and thus sees no induced signal and does not contribute to the antenna's operation. The end closer to the source is referred to as the front. Near the rear is a single active element, typically a half-wave dipole or folded dipole. Passive elements are arranged in front (*directors*) and behind (*reflectors*) the active element along the boom. The Yagi has the inherent quality that it becomes increasingly directional, and thus has higher gain, as the number of elements increases. However, this also makes it increasingly sensitive to changes in frequency; if the signal frequency changes, not only does the active element receive less energy directly, but all of the passive elements adding to that signal also decrease their output as well and their signals no longer reach the active element in-phase.

It is also possible to use multiple active elements and combine them together with transmission lines to produce a similar system where the phases add up to reinforce the output. The antenna array and very similar reflective array antenna consist of multiple elements, often half-wave dipoles, spaced out on a plane and wired together with transmission lines with specific phase lengths to produce a single in-phase signal at the output. The log-periodic antenna is a more complex design that uses multiple in-line elements similar in appearance to the Yagi-Uda but using transmission lines between the elements to produce the output.

Reflection of the original signal also occurs when it hits an extended conductive surface, in a fashion similar to a mirror. This effect can also be used to increase signal through the use of a *reflector*, normally placed behind the active element and spaced so the reflected signal reaches the element in-phase. Generally the reflector will remain highly reflective even if it is not solid; gaps less than $\frac{1}{10}$ λ generally have little effect on the outcome. For this reason, reflectors often take the form of wire meshes or rows of passive elements, which makes them lighter and less subject to wind-load effects, of particular importance when mounted at higher elevations with respect to the surrounding structures. The parabolic reflector is perhaps the best known example of a reflector-based antenna, which has an effective area far greater than the active element alone.

Bandwidth

Although a resonant antenna has a purely resistive feed-point impedance at a particular frequency, many (if not most) applications require using an antenna over a range of frequencies. The frequency range or *bandwidth* over which an antenna functions well can be very wide (as in a log-periodic antenna) or narrow (in a resonant antenna); outside this range the antenna impedance becomes a poor match to the transmission line and transmitter (or receiver). Also in the case of the Yagi-Uda and other end-fire arrays, use of the antenna well away from its design frequency affects its radiation pattern, reducing its directive gain; the usable bandwidth is then limited regardless of impedance matching.

Except for the latter concern, the resonant frequency of an antenna system can always be altered by adjusting a suitable matching network. This is most efficiently accomplished using a matching network at the site of the antenna, since simply adjusting a matching network at the transmitter (or receiver) would leave the transmission line with a poor standing wave ratio.

Instead, it is often desired to have an antenna whose impedance does not vary so greatly over a certain bandwidth. It turns out that the amount of reactance seen at the terminals of a resonant antenna when the frequency is shifted, say, by 5%, depends very much on the diameter of the conductor used. A long thin wire used as a half-wave dipole (or quarter wave monopole) will have a reactance significantly greater than the resistive impedance it has at resonance, leading to a poor match and generally unacceptable performance. Making the element using a tube of a diameter perhaps 1/50 of its length, however, results in a reactance at this altered frequency which is not so great, and a much less serious mismatch and effect on the antenna's net performance. Thus rather thick tubes are often used for the elements; these also have reduced parasitic resistance (loss).

Rather than just using a thick tube, there are similar techniques used to the same effect such as replacing thin wire elements with *cages* to simulate a thicker element. This widens the bandwidth of the resonance. On the other hand, it is desired for amateur radio antennas to operate at several bands which are widely separated from each other (but not in between). This can often be accomplished simply by connecting elements resonant at those different frequencies in parallel. Most of the transmitter's power will flow into the resonant element while the others present a high (reactive) impedance, thus drawing little current from the same voltage. Another popular solution uses so-called *traps* consisting of parallel resonant circuits which are strategically placed in breaks along each antenna element. When used at one particular frequency band the trap presents a very high impedance (parallel resonance) effectively truncating the element at that length, making it a proper resonant antenna. At a lower frequency the trap allows the full length of the element to be employed, albeit with a shifted resonant frequency due to the inclusion of the trap's net reactance at that lower frequency.

The bandwidth characteristics of a resonant antenna element can be characterized according to its Q, just as one uses to characterize the sharpness of an L-C resonant circuit. A common mistake is to assume that there is an advantage in an antenna having a high Q (the so-called "quality factor"). In the context of electronic circuitry a low Q generally signifies greater loss (due to unwanted resistance) in a resonant L-C circuit, and poorer receiver selectivity. However this understanding does not apply to resonant antennas where the resistance involved is the radiation resistance, a desired quantity which removes energy from the resonant element in order to radiate it (the purpose of an antenna, after all!). The Q of an L-C-R circuit is defined as the ratio of the inductor's (or capacitor's) reactance to the resistance, so for a certain radiation resistance (the radiation resistance at resonance does not vary greatly with diameter) the greater reactance off-resonance causes the poorer bandwidth of an antenna employing a very thin conductor. The Q of such a narrowband antenna can be as high as 15. On the other hand, the reactance at the same off-resonant frequency of one using thick elements is much less, consequently resulting in a Q as low as 5. These two antennas may perform equivalently at the resonant frequency, but the second antenna will perform over a bandwidth 3 times as wide as the antenna consisting of a thin conductor.

Antennas for use over much broader frequency ranges are achieved using further techniques. Adjustment of a matching network can, in principle, allow for any antenna to be matched at any frequency. Thus the loop antenna built into most AM broadcast (medium wave) receivers has a very narrow bandwidth, but is tuned using a parallel capacitance which is adjusted according to the receiver tuning. On the other hand, log-periodic antennas are *not* resonant at any frequency but can be built to attain similar characteristics (including feedpoint impedance) over any frequency range. These are therefore commonly used (in the form of directional log-periodic dipole arrays) as television antennas.

Gain

Gain is a parameter which measures the degree of directivity of the antenna's radiation pattern. A high-gain antenna will radiate most of its power in a particular direction, while a low-gain antenna will radiate over a wider angle. The *antenna gain*, or *power gain* of an antenna is defined as the ratio of the intensity (power per unit surface area) I radiated by the antenna in the direction of its maximum output, at an arbitrary distance, divided by the intensity I_{iso} radiated at the same distance by a hypothetical isotropic antenna which radiates equal power in all directions. This dimensionless ratio is usually expressed logarithmically in decibels, these units are called "decibels-isotropic" (dBi)

$$G_{dBi} = 10 \log \frac{I}{I_{iso}}$$

A second unit used to measure gain is the ratio of the power radiated by the antenna to the power radiated by a half-wave dipole antenna I_{dipole}; these units are called "decibels-dipole" (dBd)

$$G_{dBd} = 10 \log \frac{I}{I_{dipole}}$$

Since the gain of a half-wave dipole is 2.15 dBi and the logarithm of a product is additive, the gain in dBi is just 2.15 decibels greater than the gain in dBd

$$G_{dBi} = G_{dBd} + 2.15$$

High-gain antennas have the advantage of longer range and better signal quality, but must be aimed carefully at the other antenna. An example of a high-gain antenna is a parabolic dish such as a satellite television antenna. Low-gain antennas have shorter range, but the orientation of the antenna is relatively unimportant. An example of a low-gain antenna is the whip antenna found on portable radios and cordless phones. Antenna gain should not be confused with amplifier gain, a separate parameter measuring the increase in signal power due to an amplifying device.

Effective Area or Aperture

The *effective area* or effective aperture of a receiving antenna expresses the portion of the power of a passing electromagnetic wave which it delivers to its terminals, expressed in terms of an equivalent area. For instance, if a radio wave passing a given location has a flux of 1 pW / m² (10^{-12} watts per square meter) and an antenna has an effective area of 12 m², then the antenna would deliver 12 pW of RF power to the receiver (30 microvolts rms at 75 ohms). Since the receiving antenna is not equally sensitive to signals received from all directions, the effective area is a function of the direction to the source.

Due to reciprocity (discussed above) the gain of an antenna used for transmitting must be proportional to its effective area when used for receiving. Consider an antenna with no loss, that is, one whose electrical efficiency is 100%. It can be shown that its effective area averaged over all directions must be equal to $\lambda^2/4\pi$, the wavelength squared divided by 4π. Gain is defined such that the average gain over all directions for an antenna with 100% electrical efficiency is equal to 1. Therefore, the effective area A_{eff} in terms of the gain G in a given direction is given by:

$$A_{eff} = \frac{\lambda^2}{4\pi}G$$

For an antenna with an efficiency of less than 100%, both the effective area and gain are reduced by that same amount. Therefore, the above relationship between gain and effective area still holds. These are thus two different ways of expressing the same quantity. A_{eff} is especially convenient when computing the power that would be received by an antenna of a specified gain, as illustrated by the above example.

Radiation Pattern

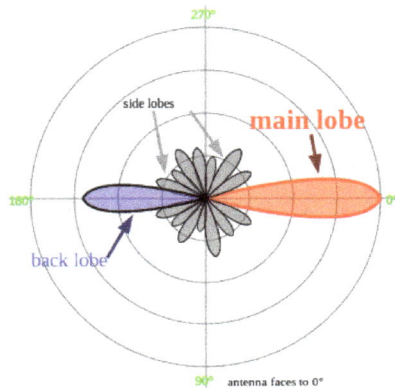

Polar plots of the horizontal cross sections of a (virtual) Yagi-Uda-antenna. Outline connects points with 3db field power compared to an ISO emitter.

The radiation pattern of an antenna is a plot of the relative field strength of the radio waves emitted by the antenna at different angles. It is typically represented by a three-dimensional graph, or polar plots of the horizontal and vertical cross sections. The pattern of an ideal isotropic antenna, which radiates equally in all directions, would look like a sphere. Many nondirectional antennas, such as monopoles and dipoles, emit equal power in all horizontal directions, with the power dropping off at higher and lower angles; this is called an omnidirectional pattern and when plotted looks like a torus or donut.

The radiation of many antennas shows a pattern of maxima or "*lobes*" at various angles, separated by "*nulls*", angles where the radiation falls to zero. This is because the radio waves emitted by different parts of the antenna typically interfere, causing maxima at angles where the radio waves arrive at distant points in phase, and zero radiation at other angles where the radio waves arrive out of phase. In a directional antenna designed to project radio waves in a particular direction, the lobe in that direction is designed larger than the others and is called the "*main lobe*". The other lobes usually represent unwanted radiation and are called "*sidelobes*". The axis through the main lobe is called the "*principal axis*" or "*boresight axis*".

Field Regions

The space surrounding an antenna can be divided into three concentric regions: the reactive near-field, the radiating near-field (Fresnel region) and the far-field (Fraunhofer) regions. These regions are useful to identify the field structure in each, although there are no precise boundaries.

The far-field region is are far enough from the antenna to neglect its size and shape. It can be assumed that the electromagnetic wave is purely a radiating plane wave (electric and magnetic fields are in phase and perpendicular to each other and to the direction of propagation). This simplifies the mathematical analysis of the radiated field.

Impedance

As an electro-magnetic wave travels through the different parts of the antenna system (radio, feed line, antenna, free space) it may encounter differences in impedance (E/H, V/I, etc.). At each interface, depending on the impedance match, some fraction of the wave's energy will reflect back to the source, forming a standing wave in the feed line. The ratio of maximum power to minimum power in the wave can be measured and is called the standing wave ratio (SWR). A SWR of 1:1 is ideal. A SWR of 1.5:1 is considered to be marginally acceptable in low power applications where power loss is more critical, although an SWR as high as 6:1 may still be usable with the right equipment. Minimizing impedance differences at each interface (impedance matching) will reduce SWR and maximize power transfer through each part of the antenna system.

Complex impedance of an antenna is related to the electrical length of the antenna at the wavelength in use. The impedance of an antenna can be matched to the feed line and radio by adjusting the impedance of the feed line, using the feed line as an impedance transformer. More commonly, the impedance is adjusted at the load with an antenna tuner, a balun, a matching transformer, matching networks composed of inductors and capacitors, or matching sections such as the gamma match.

Efficiency

Efficiency of a transmitting antenna is the ratio of power actually radiated (in all directions) to the power absorbed by the antenna terminals. The power supplied to the antenna terminals which is not radiated is converted into heat. This is usually through loss resistance in the antenna's conductors, but can also be due to dielectric or magnetic core losses in antennas (or antenna systems) using such components. Such loss effectively robs power from the transmitter, requiring a stronger transmitter in order to transmit a signal of a given strength.

For instance, if a transmitter delivers 100 W into an antenna having an efficiency of 80%, then the antenna will radiate 80 W as radio waves and produce 20 W of heat. In order to radiate 100 W of power, one would need to use a transmitter capable of supplying 125 W to the antenna. Note that antenna efficiency is a separate issue from impedance matching, which may also reduce the amount of power radiated using a given transmitter. If an SWR meter reads 150 W of incident power and 50 W of reflected power, that means that 100 W have actually been absorbed by the antenna (ignoring transmission line losses). How much of that power has actually been radiated cannot be directly determined through electrical measurements at (or before) the antenna terminals, but would require (for instance) careful measurement of field strength. Fortunately the loss resistance of antenna conductors such as aluminum rods can be calculated and the efficiency of an antenna using such materials predicted.

However loss resistance will generally affect the feedpoint impedance, adding to its resistive (real) component. That resistance will consist of the sum of the radiation resistance R_r and the loss resistance R_{loss}. If an rms current I is delivered to the terminals of an antenna, then a power of I^2R_r will be radiated and a power of I^2R_{loss} will be lost as heat. Therefore, the efficiency of an antenna is equal to $R_r / (R_r + R_{loss})$. Of course only the total resistance $R_r + R_{loss}$ can be directly measured.

According to reciprocity, the efficiency of an antenna used as a receiving antenna is identical to the efficiency as defined above. The power that an antenna will deliver to a receiver (with a proper impedance match) is reduced by the same amount. In some receiving applications, the very inefficient antennas may have little impact on performance. At low frequencies, for example, atmospheric or man-made noise can mask antenna inefficiency. For example, CCIR Rep. 258-3 indicates man-made noise in a residential setting at 40 MHz is about 28 dB above the thermal noise floor. Consequently, an antenna with a 20 dB loss (due to inefficiency) would have little impact on system noise performance. The loss within the antenna will affect the intended signal and the noise/interference identically, leading to no reduction in signal to noise ratio (SNR).

This is fortunate, since antennas at lower frequencies which are not rather large (a good fraction of a wavelength in size) are inevitably inefficient (due to the small radiation resistance R_r of small antennas). Most AM broadcast radios (except for car radios) take advantage of this principle by including a small loop antenna for reception which has an extremely poor efficiency. Using such an inefficient antenna at this low frequency (530–1650 kHz) thus has little effect on the receiver's net performance, but simply requires greater amplification by the receiver's electronics. Contrast this tiny component to the massive and very tall towers used at AM broadcast stations for transmitting at the very same frequency, where every percentage point of reduced antenna efficiency entails a substantial cost.

The definition of antenna gain or *power gain* already includes the effect of the antenna's efficiency. Therefore, if one is trying to radiate a signal toward a receiver using a transmitter of a given power, one need only compare the gain of various antennas rather than considering the efficiency as well. This is likewise true for a receiving antenna at very high (especially microwave) frequencies, where the point is to receive a signal which is strong compared to the receiver's noise temperature. However, in the case of a directional antenna used for receiving signals with the intention of *rejecting* interference from different directions, one is no longer concerned with the antenna efficiency, as discussed above. In this case, rather than quoting the antenna gain, one would be more concerned with the *directive gain* which does *not* include the effect of antenna (in)efficiency. The directive gain of an antenna can be computed from the published gain divided by the antenna's efficiency.

Polarization

The *polarization* of an antenna refers to the orientation of the electric field (E-plane) of the radio wave with respect to the Earth's surface and is determined by the physical structure of the antenna and by its orientation; note that this designation is totally distinct from the antenna's directionality. Thus, a simple straight wire antenna will have one polarization when mounted vertically, and a different polarization when mounted horizontally. As a transverse wave, the magnetic field of a radio wave is at right angles to that of the electric field, but by convention, talk of an antenna's "polarization" is understood to refer to the direction of the electric field.

Reflections generally affect polarization. For radio waves, one important reflector is the iono-sphere which can change the wave's polarization. Thus for signals received following reflection by the ionosphere (a skywave), a consistent polarization cannot be expected. For line-of-sight com-munications or ground wave propagation, horizontally or vertically polarized transmissions gen-erally remain in about the same polarization state at the receiving location. Matching the receiving antenna's polarization to that of the transmitter can make a very substantial difference in received signal strength.

Polarization is predictable from an antenna's geometry, although in some cases it is not at all ob-vious (such as for the quad antenna). An antenna's linear polarization is generally along the direc-tion (as viewed from the receiving location) of the antenna's currents when such a direction can be defined. For instance, a vertical whip antenna or Wi-Fi antenna vertically oriented will transmit and receive in the vertical polarization. Antennas with horizontal elements, such as most rooftop TV antennas in the United States, are horizontally polarized (broadcast TV in the U.S. usually uses horizontal polarization). Even when the antenna system has a vertical orientation, such as an array of horizontal dipole antennas, the polarization is in the horizontal direction corresponding to the current flow. The polarization of a commercial antenna is an essential specification.

Polarization is the sum of the E-plane orientations over time projected onto an imaginary plane perpendicular to the direction of motion of the radio wave. In the most general case, polarization is elliptical, meaning that the polarization of the radio waves varies over time. Two special cases are linear polarization (the ellipse collapses into a line) as discussed above, and circular polarization (in which the two axes of the ellipse are equal). In linear polarization the electric field of the radio wave oscillates back and forth along one direction; this can be affected by the mounting of the antenna but usually the desired direction is either horizontal or vertical polarization. In circular polarization, the electric field (and magnetic field) of the radio wave rotates at the radio frequency circularly around the axis of propagation. Circular or elliptically polarized radio waves are desig-nated as right-handed or left-handed using the "thumb in the direction of the propagation" rule. Note that for circular polarization, optical researchers use the opposite right hand rule from the one used by radio engineers.

It is best for the receiving antenna to match the polarization of the transmitted wave for optimum reception. Intermediate matchings will lose some signal strength, but not as much as a complete mismatch. A circularly polarized antenna can be used to equally well match vertical or horizontal linear polarizations. Transmission from a circularly polarized antenna received by a linearly polar-ized antenna (or vice versa) entails a 3 dB reduction in signal-to-noise ratio as the received power has thereby been cut in half.

Impedance Matching

Maximum power transfer requires matching the impedance of an antenna system (as seen looking into the transmission line) to the complex conjugate of the impedance of the receiver or transmit-ter. In the case of a transmitter, however, the desired matching impedance might not correspond to the dynamic output impedance of the transmitter as analyzed as a source impedance but rather the design value (typically 50 ohms) required for efficient and safe operation of the transmitting circuitry. The intended impedance is normally resistive but a transmitter (and some receivers) may have additional adjustments to cancel a certain amount of reactance in order to "tweak" the match. When a transmission line is used in between the antenna and the transmitter (or receiver) one generally would like an antenna system whose impedance is resistive and near the characteristic impedance of that transmission line in order to minimize the standing wave ratio (SWR) and the increase in transmission line losses it entails, in addition to supplying a good match at the transmitter or receiver itself.

Antenna tuning generally refers to cancellation of any reactance seen at the antenna terminals, leaving only a resistive impedance which might or might not be exactly the desired impedance (that of the transmission line). Although an antenna may be designed to have a purely resistive feedpoint impedance (such as a dipole 97% of a half wavelength long) this might not be exactly true at the frequency that it is eventually used at. In some cases the physical length of the antenna can be "trimmed" to obtain a pure resistance. On the other hand, the addition of a series inductance or parallel capacitance can be used to cancel a residual capacitative or inductive reactance, respectively.

In some cases this is done in a more extreme manner, not simply to cancel a small amount of residual reactance, but to resonate an antenna whose resonance frequency is quite different from the intended frequency of operation. For instance, a "whip antenna" can be made significantly shorter than 1/4 wavelength long, for practical reasons, and then resonated using a so-called loading coil. This physically large inductor at the base of the antenna has an inductive reactance which is the opposite of the capacitative reactance that such a vertical antenna has at the desired operating frequency. The result is a pure resistance seen at feedpoint of the loading coil; unfortunately that resistance is somewhat lower than would be desired to match commercial coax.

So an additional problem beyond canceling the unwanted reactance is of matching the remaining resistive impedance to the characteristic impedance of the transmission line. In principle this can always be done with a transformer, however the turns ratio of a transformer is not adjustable. A general matching network with at least two adjustments can be made to correct both components of impedance. Matching networks using discrete inductors and capacitors will have losses associated with those components, and will have power restrictions when used for transmitting. Avoiding these difficulties, commercial antennas are generally designed with fixed matching elements or feeding strategies to get an approximate match to standard coax, such as 50 or 75 ohms. Antennas based on the dipole (rather than vertical antennas) should include a balun in between the transmission line and antenna element, which may be integrated into any such matching network.

Another extreme case of impedance matching occurs when using a small loop antenna (usually, but not always, for receiving) at a relatively low frequency where it appears almost as a pure inductor. Resonating such an inductor with a capacitor at the frequency of operation not only cancels the reactance but greatly magnifies the very small radiation resistance of such a loop. This is implemented in most AM broadcast receivers, with a small ferrite loop antenna resonated by a capacitor which is varied along with the receiver tuning in order to maintain resonance over the AM broadcast band.

Antenna Types

Antennas can be classified in various ways. The list below groups together antennas under common operating principles, following the way antennas are classified in many engineering textbooks.

Isotropic: An isotropic antenna (isotropic radiator) is a *hypothetical* antenna that radiates equal signal power in all directions. It is a mathematical model that is used as the base of comparison to calculate the gain of real antennas. No real antenna can have an isotropic radiation pattern. However *approximately* isotropic antennas, constructed with multiple elements, are used in antenna testing.

The first four groups below are usually resonant antennas; when driven at their resonant frequency their elements act as resonators. Waves of current and voltage bounce back and forth between the ends, creating standing waves along the elements.

Dipole

"Rabbit ears" dipole variant for VHF television reception

Log-periodic dipole array covering 140-470 MHz

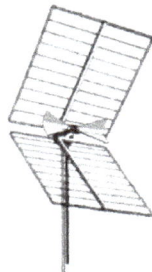

Corner reflector UHF TV antenna with "bowtie" dipole driven element

Two-element turnstile antenna for reception of weather satellite data, 137 MHz. Has circular polarization.

The dipole is the prototypical antenna on which a large class of antennas are based. A basic dipole antenna consists of two conductors (usually metal rods or wires) arranged symmetrically, with one side of the balanced feedline from the transmitter or receiver attached to each. The most common type, the half-wave dipole, consists of two resonant elements just under a quarter wavelength long. This antenna radiates maximally in directions perpendicular to the antenna's axis, giving it a small directive gain of 2.15 dBi (practically the lowest directive gain of any antenna). Although half-wave dipoles are used alone as omnidirectional antennas, they are also a building block of many other more complicated directional antennas.

- *Yagi-Uda* - One of the most common directional antennas at HF, VHF, and UHF frequencies. Consists of multiple half wave dipole elements in a line, with a single driven element and multiple parasitic elements which serve to create a uni-directional or beam antenna. These typically have gains between 10 and 20 dBi depending on the number of elements used, and are very narrowband (with a usable bandwidth of only a few percent) though there are derivative designs which relax this limitation. Used for rooftop television antennas, point-to-point communication links, and long distance shortwave communication using skywave ("skip") reflection from the ionosphere.

- *Log-periodic dipole array* - Often confused with the Yagi-Uda, this consists of many dipole elements along a boom with gradually increasing lengths, all connected to the transmission line with alternating polarity. It is a directional antenna with a wide bandwidth. This makes it ideal for use as a rooftop television antenna, although its gain is much less than a Yagi of comparable size.

- *Turnstile* - Two dipole antennas mounted at right angles, fed with a phase difference of 90°. This antenna is unusual in that it radiates in *all* directions (no nulls in the radiation pattern), with horizontal polarization in directions coplaner with the elements, circular polarization normal to that plane, and elliptical polarization in other directions. Used for receiving signals from satellites, as circular polarization is transmitted by many satellites.

- *Corner reflector* - A directive antenna with moderate gain of about 8 dBi often used at UHF frequencies. Consists of a dipole mounted in front of two reflective metal screens joined at an angle, usually 90°. Used as a rooftop UHF television antenna and for point-to-point data links.

- *Patch (microstrip)* - A type of antenna with elements consisting of metal sheets mounted over a ground plane. Similar to dipole with gain of 6 - 9 dBi. Integrated into surfaces such as aircraft bodies. Their easy fabrication using PCB techniques have made them popular in modern wireless devices. Often used in arrays.

Monopole

Quarter-wave whip antenna on an FM radio for 88-108 MHz

Rubber Ducky antenna on UHF 446 MHz walkie talkie with rubber cover removed.

Mast radiator antenna of medium wave AM radio station, Germany

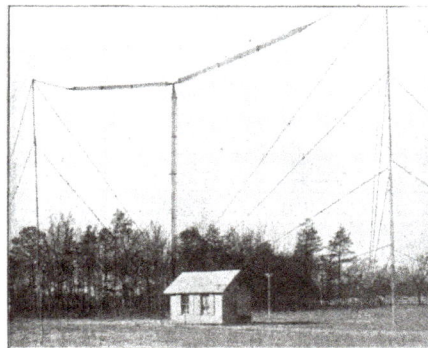

T antenna of amateur radio station, 80 ft high, used at 1.5 MHz.

Monopole antennas consist of a single conductor such as a metal rod, mounted over the ground or an artificial conducting surface (a so-called *ground plane*). One side of the feedline from the receiver or transmitter is connected to the conductor, and the other side to ground and/or the artificial ground plane. The monopole is best understood as a dipole antenna in which one conductor is omitted; the radiation is generated as if the second arm of the dipole were present due to the effective image current seen as a reflection of the monopole from the ground. Since all of the equivalent dipole's radiation is concentrated in a half-space, the antenna has twice (3 dB increase of) the gain of a similar dipole, not considering losses in the ground plane.

The most common form is the quarter-wave monopole which is one-quarter of a wavelength long and has a gain of 5.12 dBi when mounted over a ground plane. Monopoles have an omnidirectional radiation pattern, so they are used for broad coverage of an area, and have vertical polarization. The ground waves used for broadcasting at low frequencies must be vertically polarized, so large vertical monopole antennas are used for broadcasting in the MF, LF, and VLF bands. Small monopoles are used as nondirectional antennas on portable radios in the HF, VHF, and UHF bands.

- *Whip* - Type of antenna used on mobile and portable radios in the VHF and UHF bands such as boom boxes, consists of a flexible rod, often made of telescoping segments.

    o *Rubber Ducky* - Most common antenna used on portable two way radios and cordless phones due to its compactness, consists of an electrically short wire helix. The helix adds inductance to cancel the capacitive reactance of the short radiator, making it resonant. Very low gain.

    o *Ground plane* - a whip antenna with several rods extending horizontally from base of whip attached to the ground side of the feedline. Since whips are mounted above ground, the horizontal rods form an artificial ground plane under the antenna to increase its gain. Used as base station antennas for land mobile radio systems such as police, ambulance and taxi dispatchers.

- *Mast radiator* - A radio tower in which the tower structure itself serves as the antenna. Common form of transmitting antenna for AM radio stations and other MF and LF transmitters. At its base the tower is usually, but not necessarily, mounted on a ceramic insulator to isolate it from the ground.

- *T and inverted L* - Consist of a long horizontal wire suspended between two towers with insulators, with a vertical wire hanging down from it, attached to a feedline to the receiver or transmitter. Used on LF and VLF bands. The vertical wire serves as the radiator. Since at these frequencies the vertical wire is electrically short, much shorter than a quarter wavelength, the horizontal wire(s) serve as a capacitive "hat" to increase the current in the vertical radiator, increasing the gain. Very narrow bandwidth, requires loading coil to tune out the capacitive reactance and make it resonant. Requires low resistance ground (electricity).

- *Inverted F* - Combines the advantages of the inverted-L antenna and the F-type antenna of, respectively, compactness and good matching. The antenna is grounded at the base and fed at some intermediate point. The position of the feed point determines the antenna impedance. Thus, matching can be achieved without the need for an extraneous matching network.

- *Umbrella* - Very large wire transmitting antennas used on VLF bands. Consists of a central mast radiator tower attached at the top to multiple wires extending out radially from the mast to ground, like a tent or umbrella, insulated at the ends. Extremely narrow bandwidth, requires large loading coil and low resistance counterpoise ground. Used for long range military communications.

Array

Sector antennas *(white bars)* on cell phone tower. Collinear arrays of dipoles, these radiate a flat, fan-shaped beam.

108 MHz reflective array antenna of AN-270 radar used during WW2.

US Air Force PAVE PAWS phased array radar antenna for ballistic missile detection, Alaska. The two circular arrays are each composed of 2677 crossed dipole antennas.

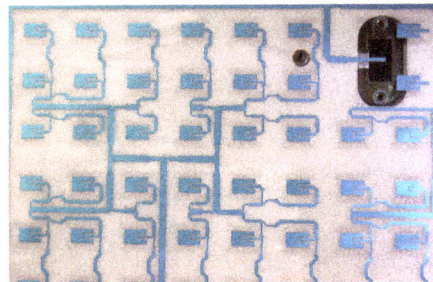

Flat microstrip array antenna for satellite TV reception.

Array antennas consist of multiple antennas working as a single antenna. Typically they consist of arrays of identical driven elements, usually dipoles fed in phase, giving increased gain over that of a single dipole.

- *Collinear* - Consist of a number of dipoles in a vertical line. It is a high gain omnidirectional antenna, meaning more of the power is radiated in horizontal directions and less into the sky or ground and wasted. Gain of 8 to 10 dBi. Used as base station antennas for land mobile radio systems such as police, fire, ambulance, and taxi dispatchers, and sector antennas for cellular base stations.

- *Reflective array* - multiple dipoles in a two-dimensional array mounted in front of a flat reflecting screen. Used for radar and UHF television transmitting and receiving antennas.

- *Phased array* - A high gain antenna used at UHF and microwave frequencies which is electronically steerable. It consists of multiple dipoles in a two-dimensional array, each fed through an electronic phase shifter, with the phase shifters controlled by a computer control system. The beam can be instantly pointed in any direction over a wide angle in front of the antenna. Used for military radar and jamming systems.

- *Curtain array* - Large directional wire transmitting antenna used at HF by shortwave broadcasting stations. It consists of a vertical rectangular array of wire dipoles suspended in front of a flat reflector screen consisting of a vertical "curtain" of parallel wires, all supported between two metal towers. It radiates a horizontal beam of radio waves into the sky above the horizon, which is reflected by the ionosphere to Earth beyond the horizon.

- *Batwing* or *superturnstile* - A specialized antenna used in television broadcasting consisting of perpendicular pairs of dipoles with radiators resembling bat wings. Multiple batwing antennas are stacked vertically on a mast to make VHF television broadcast antennas. Omnidirectional radiation pattern with high gain in horizontal directions. The batwing shape gives them wide bandwidth.

- *microstrip* - an array of patch antennas on a substrate fed by microstrip feedlines. Microwave antenna that can achieve large gains in compact space. Ease of fabrication by PCB techniques have made them popular in modern wireless devices. Beamwidth and polarization can be actively reconfigurable.

Loop

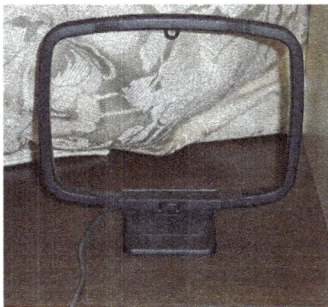

Separate loop antenna for AM radio

Loop antenna for transmitting at high frequencies, 2m diameter

Loop antennas consist of a loop (or coil) of wire. Loops with circumference of a wavelength (or integer multiple of the wavelength) are resonant and act somewhat similarly to the half-wave dipole. However a loop small in comparison to the wavelength, also called a magnetic loop, performs quite differently. This antenna interacts directly with the magnetic field of the radio wave, making it relatively insensitive to nearby electrical noise. However it has a very small radiation resistance, typically much smaller than the loss resistance, making it inefficient and thus undesirable for transmitting. They are used as receiving antennas at low frequencies, and also as direction finding antennas.

Ferrite loopstick antenna from an AM broadcast radio, about 4 in (10 cm) long. The antenna is inductive and, in conjunction with a variable capacitor, forms the tuned circuit at the input stage of the receiver.

- *Ferrite (loopstick)* - These are used as the receiving antenna in most consumer AM radios operating in the medium wave broadcast band (and lower frequencies), a notable exception being car radios. Wire is coiled around a ferrite core which greatly increases the coil's inductance. Radiation pattern is maximum at directions normal to the ferrite stick.

- *Quad* - consists of multiple wire loops in a line with one functioning as the driven element, and the others as parasitic elements. Used as a directional antenna on the HF bands for shortwave communication.

Aperture

NASA Cassegrain parabolic spacecraft communication antenna, Australia. Uses X band, 8 – 12 GHz. Extremely high gain ~70 dBi.

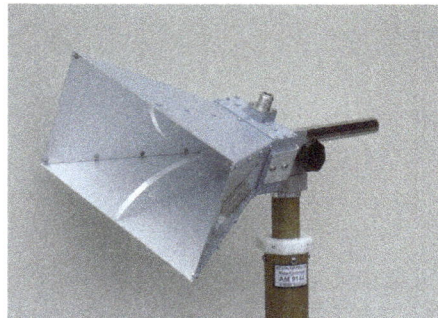

Microwave horn antenna bandwidth 0.8–18 GHz

X band marine radar slot antenna on ship, 8 – 12 GHz.

Dielectric lens antenna used in millimeter wave radio telescope

Aperture antennas are the main type of directional antennas used at microwave frequencies and above. They consist of a small dipole or loop feed antenna inside a three-dimensional guiding structure large compared to a wavelength, with an aperture to emit the radio waves. Since the antenna structure itself is nonresonant they can be used over a wide frequency range by replacing or tuning the feed antenna.

- *Parabolic* - The most widely used high gain antenna at microwave frequencies and above. Consists of a dish-shaped metal parabolic reflector with a feed antenna at the focus. It can have some of the highest gains of any antenna type, up to 60 dBi, but the dish must be large compared to a wavelength. Used for radar antennas, point-to-point data links, satellite communication, and radio telescopes.

- *Horn* - Simple antenna with moderate gains of 15 to 25 dBi consists of a flaring metal horn attached to a waveguide. Used for applications such as radar guns, radiometers and as feed antennas for parabolic dishes.

- *Slot* - Consist of a waveguide with one or more slots cut in it to emit the microwaves. Linear slot antennas emit narrow fan-shaped beams. Used as UHF broadcast antennas and marine radar antennas.

- *Dielectric resonator* - consists of small ball or puck-shaped piece of dielectric material excited by aperture in waveguide Used at millimeter wave frequencies.

Traveling Wave

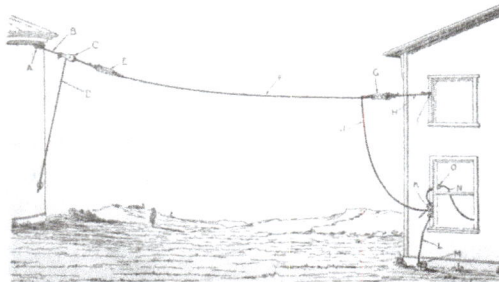

A typical random wire antenna for shortwave reception, strung between two buildings.

Unlike the above antennas, traveling wave antennas are nonresonant so they have inherently broad bandwidth. They are typically wire antennas multiple wavelengths long, through which the voltage and current waves travel in one direction, instead of bouncing back and forth to form standing waves as in resonant antennas. They have linear polarization (except for the helical antenna). Unidirectional traveling wave antennas are terminated by a resistor at one end equal to the antenna's characteristic resistance, to absorb the waves from one direction. This makes them inefficient as transmitting antennas.

Quadrant antenna, similar to rhombic, at an Austrian shortwave broadcast station.
Radiates horizontal beam at 5-9 MHz, 100 kW

Array of four axial-mode helical antennas used for satellite tracking, France

- *Random wire* - This describes the typical antenna used to receive shortwave radio, consisting of a random length of wire either strung outdoors between supports or indoors in a zig-zag pattern along walls, connected to the receiver at one end. Can have complex radiation patterns with several lobes at angles to the wire.

- *Beverage* - Simplest unidirectional traveling wave antenna. Consists of a straight wire one to several wavelengths long, suspended near the ground, connected to the receiver at one end and terminated by a resistor equal to its characteristic impedance, 400 to 800Ω at the other end. Its radiation pattern has a main lobe at a shallow angle in the sky off the terminated end. It is used for reception of skywaves reflected off the ionosphere in long distance "skip" shortwave communication.

- *Rhombic* - Consists of four equal wire sections shaped like a rhombus. It is fed by a balanced feedline at one of the acute corners, and the two sides are connected to a resistor equal to the characteristic resistance of the antenna at the other. It has a main lobe in a horizontal direction off the terminated end of the rhombus. Used for skywave communication on shortwave bands.

- *Helical (axial mode)* - Consists of a wire in the shape of a helix mounted above a reflecting screen. It radiates circularly polarized waves in a beam off the end, with a typical gain of 15 dBi. It is used at VHF and UHF frequencies. Often used for satellite communication, which uses circular polarization because it is insensitive to the relative rotation on the beam axis.

- *Leaky wave* - Microwave antennas consisting of a waveguide or coaxial cable with a slot or apertures cut in it so it radiates continuously along its length.

Effect of Ground

Ground reflections is one of the common types of multipath.

The radiation pattern and even the driving point impedance of an antenna can be influenced by the dielectric constant and especially conductivity of nearby objects. For a terrestrial antenna, the ground is usually one such object of importance. The antenna's height above the ground, as well as the electrical properties (permittivity and conductivity) of the ground, can then be important. Also, in the particular case of a monopole antenna, the ground (or an artificial ground plane) serves as the return connection for the antenna current thus having an additional effect, particularly on the impedance seen by the feed line.

When an electromagnetic wave strikes a plane surface such as the ground, part of the wave is transmitted into the ground and part of it is reflected, according to the Fresnel coefficients. If the ground is a very good conductor then almost all of the wave is reflected (180° out of phase), whereas a ground modeled as a (lossy) dielectric can absorb a large amount of the wave's power. The power remaining in the reflected wave, and the phase shift upon reflection, strongly depend on the wave's angle of incidence and polarization. The dielectric constant and conductivity (or simply the complex dielectric constant) is dependent on the soil type and is a function of frequency.

For very low frequencies to high frequencies (<30 MHz), the ground behaves as a lossy dielectric, Thus the ground is characterized both by a conductivity and permittivity (dielectric constant) which can be measured for a given soil (but is influenced by fluctuating moisture levels) or can be estimated from certain maps. At lower frequencies the ground acts mainly as a good conductor, which AM middle wave broadcast (.5 - 1.6 MHz) antennas depend on.

At frequencies between 3 and 30 MHz, a large portion of the energy from a horizontally polarized antenna reflects off the ground, with almost total reflection at the grazing angles important for ground wave propagation. That reflected wave, with its phase reversed, can either cancel or reinforce the direct wave, depending on the antenna height in wavelengths and elevation angle (for a sky wave).

On the other hand, vertically polarized radiation is not well reflected by the ground except at grazing incidence or over very highly conducting surfaces such as sea water. However the grazing angle reflection important for ground wave propagation, using vertical polarization, is *in phase* with the direct wave, providing a boost of up to 6 db, as is detailed below.

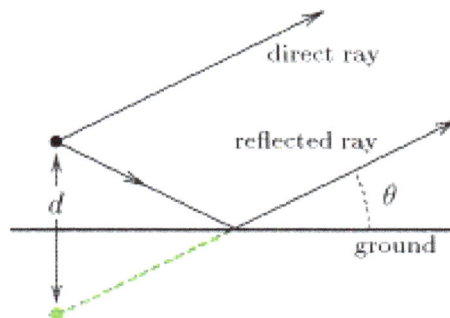

The wave reflected by earth can be considered as emitted by the image antenna.

At VHF and above (>30 MHz) the ground becomes a poorer reflector. However it remains a good reflector especially for horizontal polarization and grazing angles of incidence. That is important as these higher frequencies usually depend on horizontal line-of-sight propagation (except for satellite communications), the ground then behaving almost as a mirror.

The net quality of a ground reflection depends on the topography of the surface. When the irregularities of the surface are much smaller than the wavelength, the dominant regime is that of specular reflection, and the receiver sees both the real antenna and an image of the antenna under the ground due to reflection. But if the ground has irregularities not small compared to the wavelength, reflections will not be coherent but shifted by random phases. With shorter wavelengths (higher frequencies), this is generally the case.

Whenever both the receiving or transmitting antenna are placed at significant heights above the ground (relative to the wavelength), waves specularly reflected by the ground will travel a longer distance than direct waves, inducing a phase shift which can sometimes be significant. When a sky wave is launched by such an antenna, that phase shift is always significant unless the antenna is very close to the ground (compared to the wavelength).

The phase of reflection of electromagnetic waves depends on the polarization of the incident wave. Given the larger refractive index of the ground (typically $n=2$) compared to air ($n=1$), the phase of horizontally polarized radiation is reversed upon reflection (a phase shift of radians or 180°). On the other hand, the vertical component of the wave's electric field is reflected at grazing angles of incidence approximately *in phase*. These phase shifts apply as well to a ground modelled as a good electrical conductor.

The currents in an antenna appear as an image in *opposite* phase when reflected at grazing angles. This causes a phase reversal for waves emitted by a horizontally polarized antenna (left) but not a vertically polarized antenna (center).

This means that a receiving antenna "sees" an image of the antenna but with reversed currents. That current is in the same absolute direction as the actual antenna if the antenna is vertically oriented (and thus vertically polarized) but opposite the actual antenna if the antenna current is horizontal.

The actual antenna which is *transmitting* the original wave then also may *receive* a strong signal from its own image from the ground. This will induce an additional current in the antenna element, changing the current at the feedpoint for a given feedpoint voltage. Thus the antenna's impedance, given by the ratio of feedpoint voltage to current, is altered due to the antenna's proximity to the ground. This can be quite a significant effect when the antenna is within a wavelength or two of the ground. But as the antenna height is increased, the reduced power of the reflected wave (due to the inverse square law) allows the antenna to approach its asymptotic feedpoint impedance given by theory. At lower heights, the effect on the antenna's impedance is *very* sensitive to the exact distance from the ground, as this affects the phase of the reflected wave relative to the currents in the antenna. Changing the antenna's height by a quarter wavelength, then changes the phase of the reflection by 180°, with a completely different effect on the antenna's impedance.

The ground reflection has an important effect on the net far field radiation pattern in the vertical plane, that is, as a function of elevation angle, which is thus different between a vertically and horizontally polarized antenna. Consider an antenna at a height h above the ground, transmitting a wave considered at the elevation angle θ. For a vertically polarized transmission the magnitude of the electric field of the electromagnetic wave produced by the direct ray plus the reflected ray is:

$$|E_V| = 2|E_0|\left|\cos\left(\frac{2\pi h}{\lambda}\sin\theta\right)\right|$$

Thus the *power* received can be as high as 4 times that due to the direct wave alone (such as when $\theta=0$), following the *square* of the cosine. The sign inversion for the reflection of horizontally polarized emission instead results in:

$$|E_H| = 2|E_0|\left|\sin\left(\frac{2\pi h}{\lambda}\sin\theta\right)\right|$$

where:

E_0 is the electrical field that would be received by the direct wave if there were no ground.

θ is the elevation angle of the wave being considered.

λ is the wavelength.

h is the height of the antenna (half the distance between the antenna and its image).

Radiation patterns of antennas and their images reflected by the ground. At left the polarization is vertical and there is always a maximum for $\theta = 0$. If the polarization is horizontal as at right, there is always a zero for $\theta = 0$.

For horizontal propagation between transmitting and receiving antennas situated near the ground reasonably far from each other, the distances traveled by the direct and reflected rays are nearly the same. There is almost no relative phase shift. If the emission is polarized vertically, the two fields (direct and reflected) add and there is maximum of received signal. If the signal is polarized horizontally, the two signals subtract and the received signal is largely cancelled. The vertical plane radiation patterns are shown in the image at right. With vertical polarization there is always a maximum for $\theta=0$, horizontal propagation (left pattern). For horizontal polarization, there is cancellation at that angle. Note that the above formulae and these plots assume the ground as a perfect conductor. These plots of the radiation pattern correspond to a distance between the antenna and its image of 2.5λ. As the antenna height is increased, the number of lobes increases as well.

The difference in the above factors for the case of $\theta=0$ is the reason that most broadcasting (transmissions intended for the public) uses vertical polarization. For receivers near the ground, horizontally polarized transmissions suffer cancellation. For best reception the receiving antennas for these signals are likewise vertically polarized. In some applications where the receiving antenna must work in any position, as in mobile phones, the base station antennas use mixed polarization, such as linear polarization at an angle (with both vertical and horizontal components) or circular polarization.

On the other hand, classical (analog) television transmissions are usually horizontally polarized, because in urban areas buildings can reflect the electromagnetic waves and create ghost images due to multipath propagation. Using horizontal polarization, ghosting is reduced because the amount of reflection of electromagnetic waves in the p polarization (horizontal polarization off the side of a building) is generally less than s (vertical, in this case) polarization. Vertically polarized analog television has nevertheless been used in some rural areas. In digital terrestrial television such reflections are less problematic, due to robustness of binary transmissions and error correction.

Mutual Impedance and Interaction between Antennas

Current circulating in one antenna generally induces a voltage across the feedpoint of nearby antennas or antenna elements. The mathematics presented below are useful in analyzing the electrical behaviour of antenna arrays, where the properties of the individual array elements (such as half wave dipoles) are already known. If those elements were widely separated and driven in a certain amplitude and phase, then each would act independently as that element is known to. However, because of the mutual interaction between their electric and magnetic fields due to proximity, the currents in each element are *not* simply a function of the applied voltage (according to its driving point impedance), but depend on the currents in the other nearby elements. Note that this now is a near field phenomenon which could not be properly accounted for using the Friis transmission equation for instance. This near field effect creates a different set of currents at the antenna terminals resulting in distortions in the far field radiation patterns; however, the distortions may be removed using a simple set of network equations.

The elements' feedpoint currents and voltages can be related to each other using the concept of mutual impedance Z_{ji} between every pair of antennas just as the mutual impedance $j\omega M$ describes the voltage induced in one inductor by a current through a nearby coil coupled to it through a mutual inductance M. The mutual impedance Z_{21} between two antennas is defined as:

$$Z_{ji} = \frac{v_j}{i_i}$$

where i_i is the current flowing in antenna i and v_j is the voltage induced at the open-circuited feedpoint of antenna j due to i_1 when all other currents i_k are zero. The mutual impendances can be viewed as the elements of a symmetric square impedance matrix Z. Note that the diagonal elements, $Z_{ii} = \frac{v_i}{i_i}$, are simply the driving point impedances of each element.

Using this definition, the voltages present at the feedpoints of a set of coupled antennas can be expressed as the multiplication of the impedance matrix times the vector of currents. Written out as discrete equations, that means:

$$
\begin{aligned}
v_1 &= i_1 Z_{11} + i_2 Z_{12} + \cdots + i_n Z_{1n} \\
v_2 &= i_1 Z_{21} + i_2 Z_{22} + \cdots + i_n Z_{2n} \\
&\vdots \\
v_n &= i_1 Z_{n1} + i_2 Z_{n2} + \cdots + i_n Z_{nn}
\end{aligned}
$$

where:

v_i is the voltage at the terminals of antenna i

i_i is the current flowing between the terminals of antenna i

Z_{ii} is the driving point impedance of antenna i

Z_{ij} is the mutual impedance between antennas i and j.

As is the case for mutual inductances,

$$Z_{ij} = Z_{ji}.$$

This is a consequence of Lorentz reciprocity. For an antenna element i not connected to anything (open circuited) one can write $i_i = 0$. But for an element i which is short circuited, a current is generated across that short but no voltage is allowed, so the corresponding $v_i = 0$. This is the case, for instance, with the so-called parasitic elements of a Yagi-Uda antenna where the solid rod can be viewed as a dipole antenna shorted across its feedpoint. Parasitic elements are unpowered elements that absorb and reradiate RF energy according to the induced current calculated using such a system of equations.

With a particular geometry, it is possible for the mutual impedance between nearby antennas to be zero. This is the case, for instance, between the crossed dipoles used in the turnstile antenna.

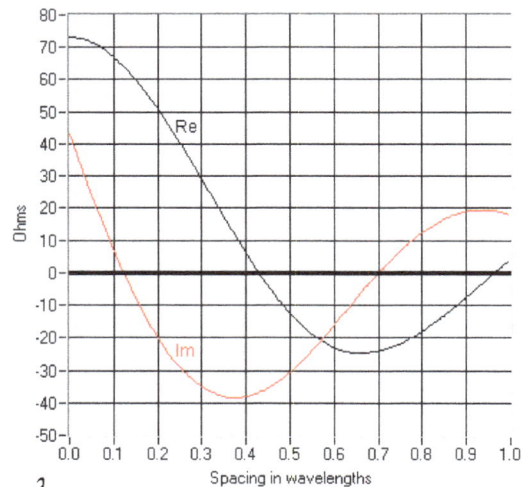

Mutual impedance between parallel $\dfrac{\lambda}{2}$ dipoles not staggered. Curves Re and Im are the resistive and reactive parts of the impedance.

Dipole Antenna

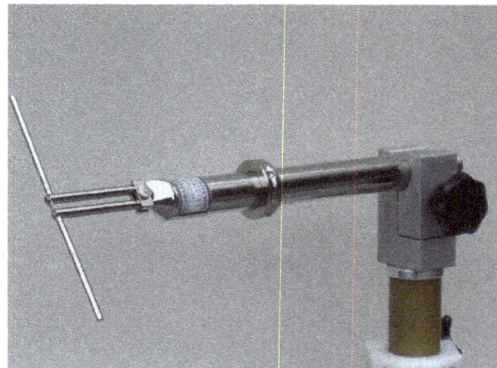

UHF Half–wave dipole

In radio and telecommunications a dipole antenna or doublet is the simplest and most widely used class of antenna. The dipole is any one of a class of antennas producing a radiation pattern approximating that of an elementary electric dipole with a radiating structure supporting a line current so energized that the current has only one node at each end. A dipole antenna commonly consists of two identical conductive elements such as metal wires or rods, which are usually bilaterally symmetrical. The driving current from the transmitter is applied, or for receiving antennas the output signal to the receiver is taken, between the two halves of the antenna. Each side of the feedline to the transmitter or receiver is connected to one of the conductors. This contrasts with a monopole antenna, which consists of a single rod or conductor with one side of the feedline connected to it, and the other side connected to some type of ground. A common example of a dipole is the "rabbit ears" television antenna found on broadcast television sets.

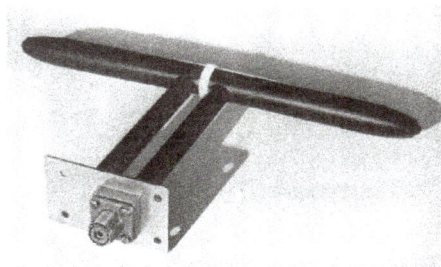

UHF half-wave dipole aircraft radar altimeter antenna

The most common form of dipole is two straight rods or wires oriented end to end on the same axis, with the feedline connected to the two adjacent ends, but dipoles may be fed anywhere along their length. This is the simplest type of antenna from a theoretical point of view. Dipoles are resonant antennas, meaning that the elements serve as resonators, with standing waves of radio current flowing back and forth between their ends. So the length of the dipole elements is determined by the wavelength of the radio waves used. The most common form is the half-wave dipole, in which each of the two rod elements is approximately 1/4 wavelength long, so the whole antenna is a half-wavelength long. The radiation pattern of a vertical dipole is omnidirectional; it radiates equal power in all azimuthal directions perpendicular to the axis of the antenna. For a half-wave dipole the radiation is maximum, 2.15 dBi perpendicular to the antenna axis, falling monotonically with elevation angle to zero on the axis, off the ends of the antenna.

Several different variations of the dipole are also used, such as the *folded dipole*, *short dipole*, *cage dipole*, *bow-tie*, and *batwing antenna*. Dipoles may be used as standalone antennas themselves, but they are also employed as feed antennas (driven elements) in many more complex antenna types, such as the Yagi antenna, parabolic antenna, reflective array, turnstile antenna, log periodic antenna, and phased array. The dipole was the earliest type of antenna; it was invented by German physicist Heinrich Hertz around 1886 in his pioneering investigations of radio waves.

Dipole Characteristics

Impedance of Dipoles of Various Lengths

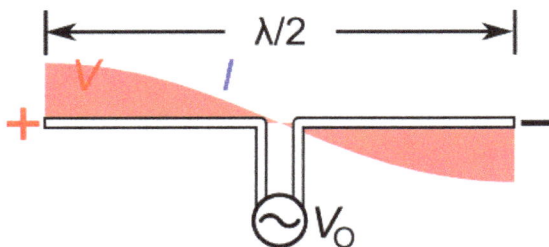

This diagram showing the sinusoidal standing waves of voltage *(V, red)* and current *(I, blue)* on a half-wave dipole driven by an AC voltage at its resonant frequency.

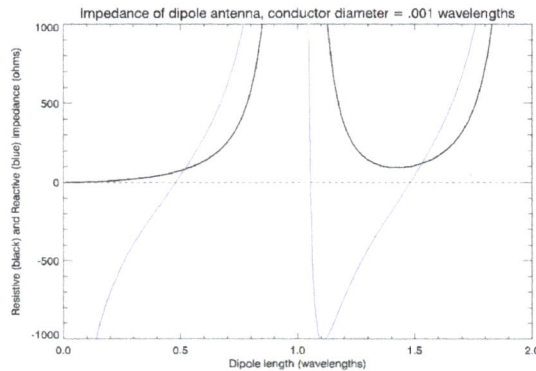

Real (black) and imaginary (blue) parts of the dipole feedpoint impedance versus total length in wavelengths, assuming a conductor diameter of .001 wavelengths

The feedpoint impedance of a dipole antenna is very sensitive to its electrical length. Therefore, a dipole will generally only perform optimally over a rather narrow bandwidth, beyond which its impedance will become a poor match for the transmitter or receiver (and transmission line). The real (resistive) and imaginary (reactive) components of that impedance, as a function of electrical length, are shown in the accompanying graph. The detailed calculation of these numbers are described below. Note that the value of the reactance is highly dependent on the diameter of the conductors; this plot is for conductors with a diameter of 0.001 wavelengths.

Dipoles that are much smaller than the wavelength of the signal are called *short dipoles*. These have a very low radiation resistance (and a high capacitive reactance) making them inefficient antennas. More of a transmitter's current is dissipated as heat due to the finite resistance of the conductors which is greater than the radiation resistance. However they can nevertheless be practical receiving antennas for longer wavelengths.

Dipoles whose length is approximately half the wavelength of the signal are called *half-wave dipoles* and are widely used as such or as the basis for derivative antenna designs. These have a radiation resistance which is much greater, closer to the characteristic impedances of available transmission lines, and normally much larger than the resistance of the conductors, so that their efficiency approaches 100%. In general radio engineering, the term *dipole*, if not further qualified, is taken to mean a center-fed half-wave dipole.

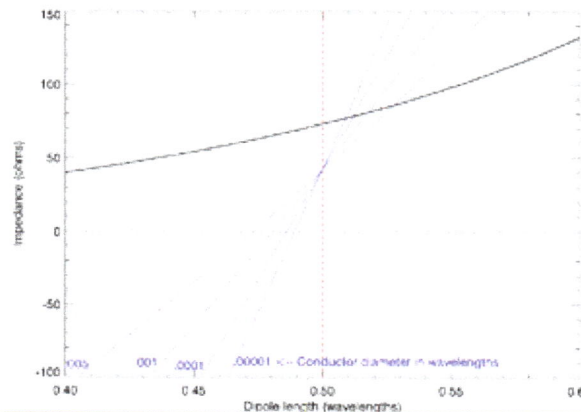

Feedpoint impedance of (near-) half-wave dipoles versus electrical length in wavelengths. Black: radiation resistance; blue: reactance for 4 different values of conductor diameter

A true half-wave dipole is one half of the wavelength λ in length, where λ=c/f in free space. Such a dipole has a feedpoint impedance consisting of 73Ω resistance and +43Ω reactance, thus presenting a slightly inductive reactance. In order to cancel that reactance, and present a pure resistance to the feedline, the element is shortened by the factor k for a net length l of:

$$l = \frac{1}{2}k\lambda = \frac{1}{2}k\frac{c}{f} = \frac{1}{2}\frac{c}{f}\frac{v}{c}$$

$$k = \frac{v}{c}$$

where λ is the free-space wavelength, c is the speed of light in free space, v is the speed of the electric wave in the wire, and f is the frequency. The adjustment factor k is equivalent to v/c. The adjustment factor k is in order for the reactance to be cancelled, depends on the diameter of the conductor. For thin wires (diameter= 0.00001 wavelengths), k is approximately 0.98; for thick conductors (diameter= 0.008 wavelengths), k drops to about 0.94. This is because the effect of antenna length on reactance is much greater for thinner conductors. For the same reason, antennas with thicker conductors have a wider operating bandwidth over which they attain an acceptable standing wave ratio.

For a typical k of about .95, the above formula is often written for a length in metres of $143/f$ or a length in feet of $468/f$ where f is the frequency in megahertz.

Dipole antennas of lengths approximately equal to any *odd* multiple of λ/2 are also resonant, presenting a small reactance (which can be cancelled by a small length adjustment). However these are rarely used. One size that is more practical though is a dipole with a length of 5/4 wavelengths. Not being close to 3/2 wavelengths, this antenna's impedance has a large (negative) reactance and can only be used with an impedance matching network (or "antenna tuner"). It is a desirable length because such an antenna has the highest gain for any dipole which isn't a great deal longer.

Radiation Pattern and Gain

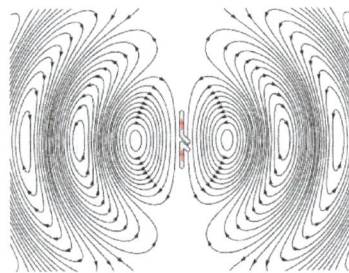

This diagram showing electric fields of a radiating vertical half-wave dipole antenna.

Three dimensional radiation pattern of a vertical half-wave dipole antenna.

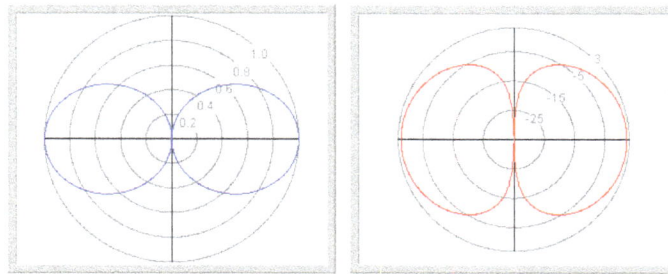

Radiation pattern of vertical half-wave dipole; vertical section. *(top)* In linear scale
(bottom) In decibels isotropic (dBi)

A dipole is omnidirectional perpendicular to the wire axis; it radiates equal power in all azimuthal directions perpendicular to the axis of its elements, with the radiation falling to zero on the axis (off the ends of the antenna). In a half wave dipole the radiation is maximum perpendicular to the antenna, declining monotonically as $(\sin \theta)^2$ to zero on the axis. Its radiation pattern in three dimensions is a toroid (doughnut) -shaped lobe symmetric about the axis of the dipole. When mounted vertically this results in maximum radiation in horizontal directions, with relatively little power radiated up into the sky or down toward the Earth, making the vertical dipole a good antenna for terrestrial communication when the direction to the receiver is unknown or changing. When mounted horizontally, the radiation pattern will have two opposing lobes at right angles (90°) to the antenna, and nulls off the ends.

Neglecting electrical inefficiency, the antenna gain is equal to the directive gain, which is 1.5 (1.76 dBi) for a short dipole, increasing to 1.64 (2.15 dBi) for a half-wave dipole. For a 5/4 wave dipole the gain further increases to about 5.2 dBi, making this length desirable for that reason even though the antenna is then off-resonance. Longer dipoles than that have radiation patterns that are multi-lobed, with poorer gain (unless they are *much* longer) even along the strongest lobe. Other enhancements to the dipole (such as including a corner reflector or an array of dipoles) can be considered when more substantial directivity is desired. Such antenna designs, although based on the half-wave dipole, generally acquire their own names.

Feeding a Dipole Antenna

Ideally, a half-wave dipole should be fed using a balanced transmission line matching its typical 65-70 Ω input impedance. Twin lead with a similar impedance is available but seldom used.

Many types of coaxial cable have a characteristic impedance of 75 Ω, which would therefore be a good match for a half-wave dipole, however, without special precautions, coax transmission line easily becomes unbalanced (with one conductor at ground potential) whereas a dipole antenna presents a balanced input (both terminals have an equal but opposite voltage with respect to ground). When an antenna is unbalanced, the unbalanced currents or "common mode" currents will flow backward along the outer conductor and the coax line will radiate, in addition to the antenna itself,. An important consequence is distortion of the antenna's designed radiation pattern, and change in the impedance seen by the line. The dipole can be properly fed, and retain its expected characteristics, by using a balun in between the coaxial feedline and the antenna terminals. Connection of coax to a dipole antenna using a balun is described in greater detail below.

Another solution, especially for receiving antennas, is to use common 300 Ω twin lead in conjunction with a *folded dipole*. The folded dipole is similar to the simple half-wave dipole but with the feedpoint impedance multiplied by 4, thus closely matching that 300 Ω impedance. This is the most common household antenna for fixed FM broadcast band tuners, which usually include balanced 300 Ω antenna input terminals.

Dipole Types

Short Dipole

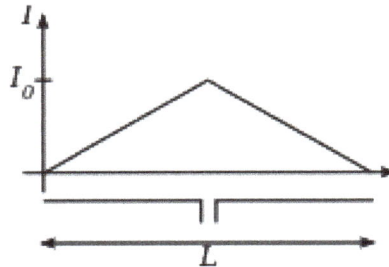

A short dipole is a dipole formed by two conductors with a total length L substantially less than a half wavelength ($\lambda/2$), the minimum length at which the antenna is resonant at the operating frequency. In order to make the antenna resonant to feed it efficiently a loading coil is required to cancel the antenna's capacitive reactance. Short dipoles are used in applications where a full half-wave dipole would be too long and cumbersome. As the length is reduced, the quantitative statements below become exact.

The feedpoint is usually at the center of the dipole. The current profile in each element, actually the tail end of a sinusoidal standing wave, is approximately a triangular distribution declining from the feedpoint current to zero at the ends. The far field electric field pattern at a distance r in the direction θ from the antenna's axis, is in the θ direction (transverse to the wave direction, in the plane of the antenna) of magnitude:

$$E_\theta = \frac{-iI_0 \sin\theta}{4\varepsilon_0 cr} \frac{L}{\lambda} e^{i(\omega t - kr)}.$$

where ω is the radian frequency ($\omega = 2\pi f$) and k is the wavenumber ($k = 2\pi/\lambda$). c is the speed of light, and the feedpoint current is assumed to be $I_0 e^{i\omega t}$.

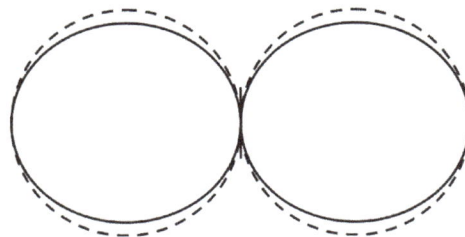

Radiation pattern of the short dipole (dashed line) compared to the half-wave dipole (solid line).

This radiation pattern is similar to and only slightly less directional than that of the half-wave dipole.

Using the above expression for the radiation in the far field for a given feedpoint current, we can integrate over all solid angle to obtain the total radiated power.

$$P_{total} = \frac{\pi}{12} I_0^2 Z_0 \left(\frac{L}{\lambda}\right)^2$$

where Z_0 is the impedance of free space, $Z_0 = 1/(c\varepsilon_0)$. From that, it is possible to infer the radiation resistance, equal to the resistive (real) part of the feedpoint impedance, neglecting a component due to ohmic losses. By setting P_{total} to the power supplied at the feedpoint $\frac{1}{2} I_0^2 R_{radiation}$ (since I_0 is the *peak* current) we find:

$$R_{radiation} = \frac{\pi}{6} Z_0 \left(\frac{L}{\lambda}\right)^2 \approx \left(\frac{L}{\lambda}\right)^2 (197\Omega).$$

Again, these relationships are accurate for L<< λ/2. Setting L=λ/2 regardless, this formula would predict a radiation resistance of 49Ω, rather than the actual 73Ω value applying to the half-wave dipole.

Full-wave Dipole

A full-wave dipole antenna consists of two half-wavelength conductors placed end to end for a total length of approximately $L = \lambda$.

The additional gain over a half-wave dipole is about 2 dB, but the impedance is much higher than a half-wave dipole making it more complicated to match with ordinary, low impedance RF equipment and cabling.

Half-wave Dipole

A half-wave dipole antenna consists of two quarter-wavelength conductors placed end to end for a total length of approximately $L = \lambda/2$.

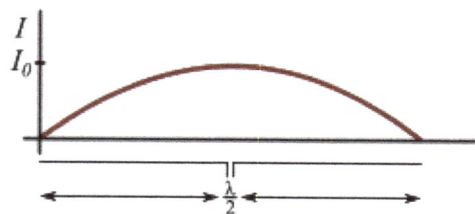

The magnitude of current in a standing wave along the dipole

The current distribution is that of a standing wave, approximately sinusoidal along the length of the dipole, with a node at each end and an antinode (peak current) at the center (feedpoint):

$$I(z) = I_0 e^{i\omega t} \cos kz,$$

where $k = 2\pi/\lambda$ and z runs from $-L/2$ to $L/2$.

In the far field, this produces a radiation pattern whose electric field is given by

$$E_\theta = \frac{-iZ_0 I_0}{2\pi r} \frac{\cos\left(\dfrac{\pi}{2}\cos\theta\right)}{\sin\theta} e^{i(\omega t - kr)}.$$

The directional factor $\cos[(\pi/2)\cos\theta]/\sin\theta$ is barely different from $\sin\theta$ applying to the short dipole, resulting in a very similar radiation pattern as noted above.

A numerical integration of this integral over all solid angle, as we did for the short dipole, supplies a value for the radiation resistance: $R_{radiation} \approx 73.1\,\Omega$. Using the induced EMF method, the real part of the driving point impedance can also be written in terms of the cosine integral:

$$R_{radiation} = \frac{Z_0}{4\pi}\text{Cin}(2\pi) = \frac{Z_0}{4\pi}\int_0^{2\pi}\frac{1-\cos(\theta)}{\theta}d\theta \approx 73.1\,\Omega.$$

If the dipole is not driven at the center, then the feed point resistance will be higher. If the feed point is distance x from one end of a half wave ($\lambda/2$) dipole, the radiation resistance relative to the feedpoint will be given by the following equation.

$$R_{radiation} = \frac{73.1\,\Omega}{\sin^2(kx)}$$

Comparing the radiated power at $\theta=0$ to the total power found by integrating, we find the directive gain to be 1.64. This can also be directly computed using the cosine integral:

$$G = \frac{4}{\text{Cin}(2\pi)} \approx 1.64 \quad (2.15\ \text{dBi})$$

Gain of dipole antennas		
length **L** in λ	Gain	Gain(dBi)
$\ll 0.5$	1.50	1.76
0.5	1.64	2.15
1.25	3.3	5.2

Quarter-wave Monopole

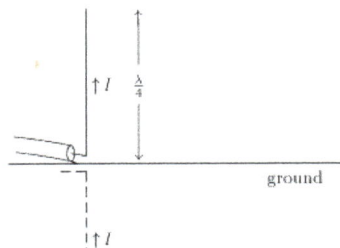

The antenna and its image form a $\dfrac{\lambda}{2}$ dipole that radiates only in the upper half of space.

The quarter-wave monopole antenna is a single-element antenna fed at one end, that behaves as a dipole antenna. It is formed by a conductor $\frac{\lambda}{4}$ in length, fed in the lower end, which is near a conductive surface which works as a reflector and is an example of a Marconi antenna. The current in the reflected image has the same direction and phase as the current in the real antenna. The quarter-wave conductor and its image together form a half-wave dipole that radiates only in the upper half of space.

In this upper side of space, the emitted field has the same amplitude of the field radiated by a half-wave dipole fed with the same current. Therefore, the total emitted power is half the emitted power of a half-wave dipole fed with the same current. As the current is the same, the radiation resistance (real part of series impedance) will be half of the series impedance of a half-wave dipole. As the reactive part is also divided by 2, the impedance of a quarter-wave antenna is $\frac{73+i43}{2}=36+i21$ ohms. Since the fields above ground are the same as for the dipole, but only half the power is applied, the gain is twice (3 dB over) that of a half-wave dipole ($\frac{\lambda}{2}$), that is, 5.14 dBi.

The earth can be used as ground plane, but it is a poor conductor. The reflected antenna image is only clear at glancing angles (far from the antenna). At these glancing angles, electromagnetic fields and radiation patterns are the same as for a half-wave dipole. Naturally, the impedance of the earth is far inferior to that of a good conductor ground plane. This can be improved (at cost) by laying a copper mesh.

When the ground is not available (such as in a vehicle) other metallic surfaces can serve as a ground plane (typically the vehicle's roof). Alternatively, radial wires placed at the base of the antenna can form a ground plane. For VHF and UHF bands, the radiating and ground plane elements can be constructed from rigid rods or tubes. For a simple 1/4-wave whip, the radials are often sloped at a 45 degree angle to bring the feed point impedance closer to 50 ohms. Since this will introduce RF energy on the shield of the unbalanced feed line which deforms the radiation pattern of the antenna, a choke is often placed near the feed point.

Folded Dipole

Folded dipole antenna

A folded dipole is a half-wave dipole with an additional wire connecting its two ends. If the additional wire has the same diameter and cross-section as the dipole, two nearly identical radiating currents are generated. The resulting far-field emission pattern is nearly identical to the one for the single-wire dipole described above, but at resonance its feedpoint impedance R_{fd} is four times the radiation resistance of a single-wire dipole. This is because for a fixed amount of power, the total radiating current I_0 is equal to twice the current in each wire and thus equal to twice the current at the feed point. Equating the average radiated power to the average power delivered at the feedpoint, we may write

$$\frac{1}{2} R_{\frac{\lambda}{2}} I_0^2 = \frac{1}{2} R_{fd} \left(I_0 / 2 \right)^2.$$

It follows that

$$R_{fd} = 4 R_{\frac{\lambda}{2}} \approx 292.32 \ \Omega.$$

The folded dipole is therefore well matched to 300 ohm balanced transmission lines, such as twin-feed ribbon cable. The folded dipole has a wider bandwidth than a single dipole. They can be used for transforming the value of input impedance of the dipole over a broad range of step-up ratios by changing the thicknesses of the wire conductors for the fed- and folded-sides. Instead of altering thickness or spacing, one can add a third parallel wire to increase the antenna impedance 9 times over a single-wire dipole, raising the impedance to 450 ohms, making a good match for window line feed cable, and further broadening the resonant frequency band of the antenna.

Half wave folded dipoles are often used for FM radio antennas; versions made with twin lead which can be hung on an inside wall often come with FM tuners. The T2FD antenna is a folded dipole. They are also widely used as driven elements for rooftop Yagi television antennas.

Other Dipole Antenna Types

There are numerous notable variations of dipole antennas:

- The *bow-tie antenna* is a dipole with flaring, triangular shaped arms. The shape gives it a much wider bandwidth than an ordinary dipole. It is widely used in UHF television antennas.

Cage dipole antennas in the Ukrainian UTR-2 radio telescope. The 8 m by 1.8 m diameter galvanized steel wire dipoles have a bandwidth of 8 - 33 MHz.

- The *cage dipole* is a similar modification in which the bandwidth is increased by using fat cylindrical dipole elements made of a "cage" of wires. These are used in a few broadband array antennas in the medium wave and shortwave bands for applications such as OTH radar and radio telescopes.

- The *vee* or *quadrant* antenna is a horizontal dipole with its arms at an angle instead of parallel. Quadrant antennas are notable in that they can be used to make horizontally polarized antennas with near-omnidirectional radiation patterns. They are used for transmitting on the HF band.

- The G5RV Antenna is a dipole antenna with a symmetric feeder line, which also serves as a 1:1 impedance transformer allowing the transceiver to see the impedance of the antenna (it does not match the antenna to the 50-ohm transceiver. In fact the impedance will be somewhere around 90 ohms at the resonant frequency but significantly different at other frequencies).

- The sloper antenna is a slanted dipole antenna used for long-range communications or in limited space.

- The AS-2259 Antenna is an inverted-V dipole antenna used for NVIS communications.

Common Applications

"Rabbit Ears" TV antenna

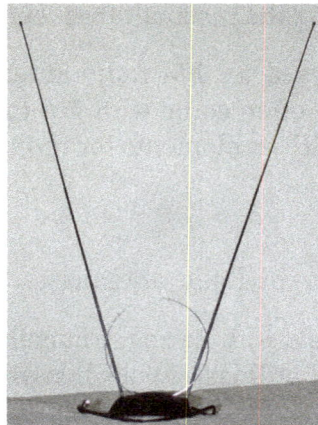

"Rabbit-ears" VHF television antenna (the small loop is a separate UHF antenna).

One of the most common applications of the dipole antenna is the *rabbit ears* or *bunny ears* television antenna, found atop broadcast television receivers. It is used to receive the VHF terrestrial television bands, consisting in the US of 52 to 88 MHz (band I) and 174 to 216 MHz (band III), with wavelengths of 5.5 to 1.4 m. Since this frequency range is much wider than a single fixed dipole antenna can cover, it is made with several degrees of adjustment. It is constructed of two telescoping rods that can be extended out to about 1 m length (approximately one quarter wavelength at 52 MHz). Instead of being fixed in opposing directions, these elements can be adjusted at an angle in a "V" shape. The reason for the V shape is that when receiving channels at the top of the band, the antenna elements will typically resonate at their 3rd harmonic. In this mode the direction of maximum gain is no longer perpendicular to the rods, but the radiation pattern will have lobes at an angle to the rods, making it advantageous to be able to adjust them to various angles.

FM Broadcast Receiving Antennas

In contrast to the wide television frequency bands, the FM broadcast band (88-108 MHz) is narrow enough that a dipole antenna can cover it. For fixed use in homes, hi-fi tuners are typically supplied with simple folded dipoles resonant near the center of that band. The feedpoint impedance of a folded dipole, which is quadruple the impedance of a simple dipole, is a good match for 300Ω twin lead, so that is usually used for the transmission line to the tuner. A common construction is to make the arms of the folded dipole out of twin lead also, shorted at their ends. This flexible antenna can be conveniently taped or nailed to walls, following the contours of mouldings.

Shortwave Antenna

Horizontal wire dipole antennas are popular for use on the HF shortwave bands, both for transmitting and shortwave listening. They are usually constructed of two lengths of wire joined by a strain insulator in the center, which is the feedpoint. The ends can be attached to existing buildings, structures, or trees, taking advantage of their heights. If used for transmitting, it is essential that the ends of the antenna be attached to supports through strain insulators with a sufficiently high flashover voltage, since the antenna's high voltage antinodes occur there. Being a balanced antenna, they are best fed with a balun between the (coax) transmission line and the feedpoint.

These are simple to put up for temporary or field use. But they are also widely used by radio amateurs and short wave listeners in fixed locations due to their simple (and inexpensive) construction, while still realizing a resonant antenna at frequencies where resonant antenna elements need to be of quite some size. They are an attractive solution for these frequencies when significant directionality is not desired, and the cost of several such resonant antennas for different frequency bands, built at home, may still be much less than a single commercially produced antenna.

Dipole Towers

Antennas for MF and LF radio stations are usually constructed as mast radiators, in which the vertical mast itself forms the antenna. Although mast radiators are most commonly monopoles, some are dipoles. The metal structure of the mast is divided at its midpoint into two insulated sections to make a vertical dipole, which is driven at the midpoint.

Dipole Arrays

Many types of array antennas are constructed using multiple dipoles, usually half-wave dipoles. The purpose of using multiple dipoles is to increase the directional gain of the antenna over the gain of a single dipole; the radiation of the separate dipoles interferes to enhance power radiated in desired directions. In arrays with multiple dipole driven elements, the feedline is split using an electrical network in order to provide power to the elements, with careful attention paid to the relative phase delays due to transmission between the common point and each element.

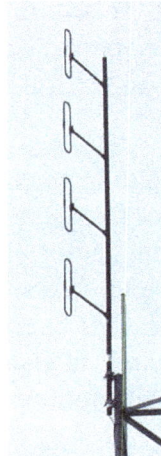

Collinear folded dipole array

For a vertically oriented dipole, which has an omnidirectional radiation pattern in the horizontal plane, it is possible to stack dipoles end-to-end fed in phase, creating a collinear antenna array. The array still has an omnidirectional pattern, but more power is radiated in the desired horizontal directions and less at large angles up into the sky or down toward the Earth. Collinear arrays are used in the VHF and UHF frequency bands at which the size of the dipoles are small enough so several can be stacked on a mast. They are a practical and higher-gain alternative to quarter wave ground plane antennas used in fixed base stations for mobile two-way radios, such as police, fire, and taxi dispatchers.

A reflective array antenna for radar consisting of numerous dipoles fed in-phase (thus realizing a *broadside array*) in front of a large reflector (horizontal wires) to make it uni-directional.

On the other hand, an array of dipoles can be used to realize substantial directivity in a particular horizontal direction. In a *broadside array* the dipoles can again be arranged colinear (end to end), or side by side, or both. The antennas are then fed in the same phase. This creates greater gain in the direction perpendicular to the antennas, at the expense of most other directions. Unfortunately that also means that the direction *opposite* the desired direction also has a high gain, whereas high gain is usually desired in one single direction. The power which is wasted in the reverse direction, however, can be recovered using a large planar reflector, as is accomplished in the reflective array antenna, increasing the gain in the desired direction by another 3 dB.

This large reflector can be avoided in the *end-fire array*. In this case the dipoles are again side by side, but are fed in different phases. Rather than being directive perpendicular to the line connecting their feedpoints, now the directivity is *along* the line connecting their feedpoints. By using an appropriate spacing and phasing, the radiation can be directed in a single direction along that line, with radiation mainly cancelled in the reverse direction as well as most other directions.

Yagi Antennas

The above described antennas with multiple driven elements require a complex feed system of signal splitting, phasing, distribution to the elements, and impedance matching. A different sort of end-fire array which is much more often used is based on the use of so-called *parasitic elements*. In the popular high-gain Yagi antenna, only one of the dipoles is actually connected electrically, but the others receive and reradiate power supplied by the driven element. This time, the phasing is accomplished by careful choice of the lengths as well as positions of the parasitic elements, in order to concentrate gain in one direction and largely cancel radiation in the opposite direction (as well as all other directions). Although the realized gain is less than a driven array with the same number of elements, the simplicity of the electrical connections makes the Yagi more practical for consumer applications.

Hertzian Dipole

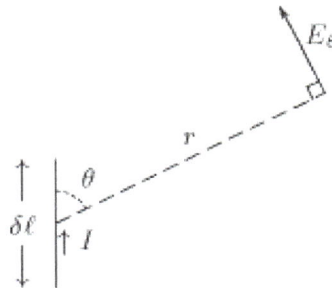

Hertzian dipole of tiny length $\delta\ell$, with current I, and field sensed at a distance r in the θ direction.

The *Hertzian dipole* or *Elementary doublet* refers to a theoretical construction, rather than a physical antenna design. It may be defined as a finite oscillating current (in a specified direction) of $Ie^{i\omega t}$ over a tiny or infinitesimal length $\delta\ell$ at a specified position. The solution of the fields from a Hertzian dipole can be used as the basis for analytical or numerical calculation of the radiation from more complex antenna geometries (such as practical dipoles) by forming the superposition of fields from a large number of Hertzian dipoles comprising the current pattern of the actual antenna. As a function of position, taking the elementary current elements multiplied by infinitesimal lengths $I(\mathbf{r})d\ell$, the resulting field pattern then reduces to an integral over the path of an antenna conductor (modelled as a thin wire).

For the following derivation we shall take the current to be in the Z direction centered at the origin ($x=y=z=0$), with the sinusoidal time dependence $e^{i\omega t}$ for all quantities being understood. The simplest approach is to use the calculation of the vector potential A(r) using the formula for the retarded potential. Although the value of A is not unique, we shall constrain it according to the Lorenz gauge, and assuming sinusoidal current at radian frequency ω the retardation of the field is converted just into a phase factor e^{-ikr}, where the wavenumber $k = \omega/c$ in free space and r is the distance between the point being considered to the origin (where we assumed the current source to be), thus $r = |\mathbf{r}|$. This results in a vector potential A at position r due to that current element only, which we find is purely in the Z direction (the direction of the current):

$$A(r) = I\delta\ell\frac{\mu_0}{4\pi r}e^{-ikr}\hat{z}$$

where μ_0 is the permeability of free space. Then using

$$\mu H = B = \nabla \times A$$

we can solve for the magnetic field H, and from that (dependent on us having chosen the Lorenz gauge) the electric field E using

$$E = \frac{\nabla \times H}{i\omega\epsilon}.$$

In spherical coordinates we find that the magnetic field H has only a component in the φ direction:

$$H_\phi = i\frac{I\delta\ell}{4\pi}\left(\frac{k}{r} - \frac{i}{r^2}\right)e^{-ikr}\sin(\theta)$$

while the electric field has components both in the θ and **r** directions:

$$E_\theta = i\frac{Z_0 I\delta\ell}{4\pi}\left(\frac{k}{r} - \frac{i}{r^2} - \frac{1}{kr^3}\right)e^{-ikr}\sin(\theta)$$

$$E_r = \frac{Z_0 I\delta\ell}{2\pi}\left(\frac{1}{r^2} - \frac{i}{kr^3}\right)e^{-ikr}\cos(\theta)$$

where $Z_0 = \sqrt{\mu_0/\varepsilon_0}$ is the impedance of free space.

This solution includes near field terms which are very strong near the source but which are *not* radiated. As seen above, the E and H fields very close to the source are almost 90° out of phase, thus contributing very little to the Poynting vector by which radiated flux is computed. The near field solution for an antenna element (from the integral using this formula over the length of that element) is the field that can be used to compute the mutual impedance between it and another nearby element.

For computation of the far field radiation pattern, the above equations are simplified as only the $1/r$ terms remain significant:

$$H_\phi = i\frac{I\delta\ell k}{4\pi r}e^{-ikr}\sin(\theta)$$

$$E_\theta = i\frac{Z_0 I\delta\ell k}{4\pi r}e^{-ikr}\sin(\theta).$$

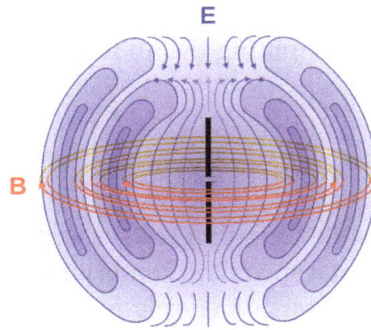

Electric field lines *(blue)* and magnetic field components *(red)* at right angles composing
the electromagnetic wave radiated by the current element (black).

The far field pattern is thus seen to consist of a transverse electromagnetic (TEM) wave, with electric and magnetic fields at right angles to each other and at right angles to the direction of propagation (the direction of r, as we assumed the source to be at the origin). The electric polarization, in the θ direction, is coplanar with the source current (in the Z direction), while the magnetic field is at right angles to that, in the φ direction. It can be seen from these equations, and also in the diagram, that the fields at these distances are exactly *in phase*. Both fields fall according to $1/r$, with the power thus falling according to $1/r^2$ as dictated by the inverse square law.

Radiation Resistance

If one knows the far field radiation pattern due to a given antenna current, then it is possible to compute the radiation resistance directly. For the above fields due to the Hertzian dipole, we can compute the power flux according to the Poynting vector, resulting in a power (as averaged over one cycle) of:

$$\langle S \rangle = \frac{1}{2} \mathrm{Re}\left(E \times H^* \right).$$

Although not required, it is simplest to do the following exercise at a large r where the far field expressions for E and H apply. Consider a large sphere surrounding the source with a radius r. We find the power per unit area crossing the surface of that sphere to be in the direction according to :

$$\langle S_r \rangle = \frac{Z_0}{2} \frac{k^2 |I|^2 (\delta \ell)^2}{(4\pi r)^2} \sin^2 \theta$$

Integration of this flux over the complete sphere results in:

$$P_{net} = \int_0^{2\pi} \int_0^{\pi} \langle S_r \rangle r^2 \sin \theta \, d\phi \, d\theta$$

$$= \frac{Z_0}{12\pi} k^2 |I|^2 (\delta \ell)^2 = \frac{\pi Z_0}{3} |I|^2 \left(\frac{\delta \ell}{\lambda} \right)^2$$

where $\lambda = 2\pi / k$ is the free space wavelength corresponding to the radian frequency ω. By definition, the radiation resistance R_{rad} times the average of the square of the current $|I|^2/2$ is the net power radiated due to that current, so equating the above to $|I|^2 R_{rad}/2$ we find:

$$R_{rad} = \frac{2\pi}{3} Z_0 \left(\frac{\delta\ell}{\lambda} \right)^2.$$

This method can be used to compute the radiation resistance for any antenna whose far field radiation pattern has been found in terms of a specific antenna current. If ohmic losses in the conductors are neglected, the radiation resistance (considered relative to the feedpoint) is identical to the resistive (real) component of the feedpoint impedance. Unfortunately this exercise tells us nothing about the reactive (imaginary) component of feedpoint impedance, whose calculation is considered below.

Directive Gain

Using the above expression for the radiated flux given by the Poynting vector, it is also possible to compute the directive gain of the Hertzian dipole. Dividing the total power computed above by $4\pi r^2$ we can find the flux averaged over all directions P_{avg} as

$$P_{avg} = \frac{P_{net}}{4\pi r^2} = \frac{Z_0}{48\pi^2 r^2} k^2 |I|^2 (\delta\ell)^2.$$

Dividing the flux radiated in a *particular* direction by P_{avg} we obtain the directive gain $G(\theta)$:

$$G(\theta) = \frac{\langle \mathbf{S}_r \rangle}{P_{avg}} = \frac{3}{2} \sin^2 \theta.$$

The commonly quoted antenna "gain", meaning the peak value of the gain pattern (radiation pattern), is found to be 1.5 or 1.76 dBi, lower than practically any other antenna configuration.

Comparison with the Short Dipole

The Hertzian dipole is *similar to* but differs from the short dipole, discussed above. In both cases the conductor is very short compared to a wavelength, so the standing wave pattern present on a half wave dipole (for instance) is absent. However, with the Hertzian dipole we specified that the current along that conductor is *constant* over its short length. This makes the Hertzian dipole useful for analysis of more complex antenna configurations, where every infinitesimal section of that *real* antenna's conductor can be modelled as a Hertzian dipole with the current found to be flowing in that real antenna.

However a short conductor fed with a RF voltage will *not* have a uniform current even along that short range. Rather, a short dipole in real life has a current equal to the feedpoint current at the feedpoint but falling linearly to zero over the length of that short conductor. By placing a *capacitive hat*, such as a metallic ball, at the end of the conductor, it is possible for its self capacitance to absorb the current from the conductor and better approximate the constant current assumed for the Hertzian dipole. But again, the Hertzian dipole is meant only as a theoretical construct for antenna analysis.

The short dipole, with a feedpoint current of I_o, has an *average* current over each conductor of only $I_o/2$. The above field equations for the Hertzian dipole of length $\delta\ell$ would then predict the *actual* fields for a short dipole using that effective current $I = I_o/2$. This would result in a power measured in the far field of *one quarter* that given by the above equation for the Poynting vector $\langle S_r \rangle$ if we had assumed an element current of I_o. Consequently, it can be seen that the radiation resistance computed for the short dipole is one quarter of that computed above for the Hertzian dipole. But their radiation patterns (and gains) are identical.

Detailed Calculation of Dipole Feedpoint Impedance

The impedance seen at the feedpoint of a dipole of various lengths has been plotted above, in terms of the real (resistive) component R_{dipole} and the imaginary (reactive) component jX_{dipole} of that impedance. For the case of an antenna with perfect conductors (no ohmic loss), R_{dipole} is identical to the radiation resistance, which can more easily be computed from the total power in the far-field radiation pattern for a given applied current as we showed for the short dipole. The calculation of X_{dipole} is more difficult.

Induced EMF Method

Using the *induced EMF method* closed form expressions are obtained for both components of the feedpoint impedance; such results are plotted above. The solution depends on an assumption for the form of the current distribution along the antenna conductors. For wavelength to element diameter ratios greater than about 60, the current distribution along each antenna element is very well approximated as a sine wave along each conductor:

$$I(z) = A\sin(k(L/2-|z|))$$

where L is the full length of the dipole, z is the position along the dipole relative to the feedpoint, k is the wavenumber equal to $2\pi/\lambda$ (λ being the wavelength, $\lambda=c/f$ for an antenna in free space), and A is an amplitude chosen to match an assumed driving point current by setting $z=0$.

In cases where an approximately sinusoidal current distribution can be assumed, this method solves for the driving point impedance in closed form using the cosine and sine integral functions Si(x) and Ci(x). For a dipole of total length L using conductors with a radius a operating at a frequency with wavenumber k ($k = 2\pi f/c$ in free space) in a medium with characteristic impedance Z_m (usually Z_o with the antenna in free space), then the resistance R and reactance X of the driving point impedance can be expressed as:

$$R_{dipole} = \frac{Z_m}{2\pi\sin^2(kL/2)}\{\gamma + \ln(kL) - Ci(kL) + \frac{1}{2}\sin(kL)[Si(2kL) - 2Si(kL)]$$

$$+ \frac{1}{2}\cos(kL)[\gamma + \ln(kL/2) + Ci(2kL) - 2Ci(kL)]\}$$

$$X_{dipole} = \frac{Z_m}{4\pi\sin^2(kL/2)}\{2Si(kL) + \cos(kL)[2Si(kL) - Si(2kL)]$$

$$- \sin(kL)[2Ci(kL) - Ci(2kL) - Ci(2ka^2/L)]\},$$

where γ is the Euler constant.

This computation using the induced EMF method is identical to the computation of the mutual impedance between two dipoles (with infinitesimal conductor radius) separated by the distance a. Because the field at or beyond the edge of an antenna's cylindrical conductor at a distance a is only dependent on the current distribution along the conductor, and not the radius of the conductor, that field is used to compute the mutual impedance between that filamentary antenna and the *actual* position of the conductor with a radius a. This then supplies the self-impedance of the conductor itself.

Integral Methods

Note that the induced EMF method is dependent on the assumption of a sinusoidal current distribution, delivering an accuracy better than about 10% as long as the wavelength to element diameter ratio is greater than about 60. However, for yet larger conductors numerical solutions are required which solve for the conductor's current distribution (rather than *assuming* a sinusoidal pattern). This can be based on approximating solutions for either *Pocklington's integrodifferential equation* or the *Hallén integral equation*. These approaches also have greater generality, not being limited to linear conductors.

Numerical solution of either is performed using the *moment method solution* which requires expansion of that current into a set of basis functions; one simple (but not the best) choice, for instance, is to break up the conductor into N segments with a constant current assumed along each. After setting an appropriate weighting function the cost may be minimized through the inversion of a NxN matrix. Determination of each matrix element requires at least one double integration involving the weighting functions, which may become computationally intensive. These are simplified if the weighting functions are simply delta functions, which corresponds to fitting the boundary conditions for the current along the conductor at only N discrete points. Then the NxN matrix must be inverted, which is also computationally intensive as N increases. In one simple example, *Balanis* performs this computation to find the antenna impedance with different N using Pocklington's method and finds that with $N>60$ solutions have approached their limiting values to within a few percent.

Feeding a Dipole using a Balun

Feeding a Dipole Antenna with Coax Cable

A dipole is a symmetrical antenna, as it is composed of two symmetrical ungrounded elements. Therefore, it works best when fed by a balanced transmission line, such as a ladder line, because in that case the symmetry (one aspect of the impedance complex, which is a complex number) matches and therefore the power transfer is optimum.

When a dipole with an unbalanced feedline such as coaxial cable is used for transmitting, the shield side of the cable, in addition to the antenna, radiates. This can induce radio frequency (RF) currents into other electronic equipment near the radiating feedline, causing RF interference. Furthermore, the antenna is not as efficient as it could be because it is radiating closer to the ground and its radiation pattern may be asymmetrically distorted. To prevent this, dipoles fed by coaxial cables have a balun between the cable and the antenna, to convert the unbalanced signal provided by the coax to a balanced symmetrical signal for the antenna.

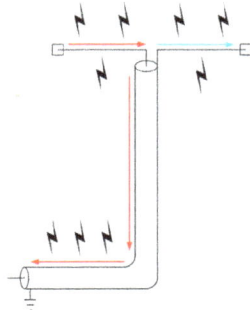

Coax and antenna both acting as radiators instead of only the antenna.

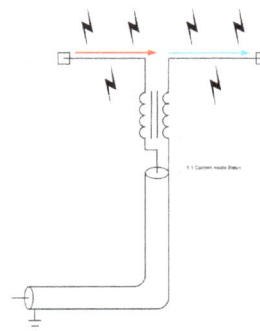

Dipole with a current balun.

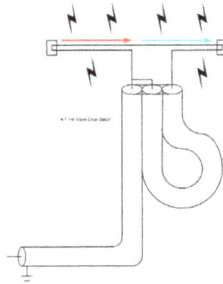

A folded dipole (300 Ω) to coax (75 Ω) 4:1 balun.

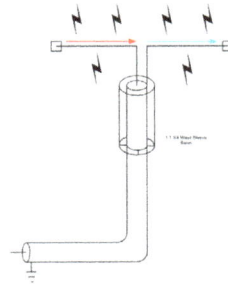

Dipole using a sleeve balun.

Several types of balun are commonly used to feed a dipole antenna: current baluns and coax baluns. Baluns can be made using ferrite toroid cores or even from the coax feedline itself. The choice of the toroid core is crucial. A rule of thumb is: the more power, the bigger the core.

Current Balun

A current balun consists of two windings that are closely coupled.

Coax Balun

A coax balun is a cost-effective method of eliminating feeder radiation, but is limited to a narrow set of operating frequencies.

One easy way to make a balun is to use a length of coaxial cable equal to half a wavelength. The inner core of the cable is linked at each end to one of the balanced connections for a feeder or dipole. One of these terminals should be connected to the inner core of the coaxial feeder. All three braids should be connected together. This then forms a 4:1 balun, which works correctly at only a narrow band of frequencies.

Sleeve Balun

At VHF frequencies, a sleeve balun can also be built to remove feeder radiation.

Another narrow-band design is to use a $\lambda/4$ length of metal pipe. The coaxial cable is placed inside the pipe; at one end the braid is wired to the pipe while at the other end no connection is made to the pipe. The balanced end of this balun is at the end where no connection is made to the pipe. The $\lambda/4$ conductor acts as a transformer, converting the zero impedance at the short to the braid into an infinite impedance at the open end. This infinite impedance at the open end of the pipe prevents current flowing into the outer coax formed by the outside of the inner coax shield and the pipe, forcing the current to remain in the inside coax. This balun design is impractical for low frequencies because of the long length of pipe that will be needed.

Dipole as a Reference Standard

Antenna gain is sometimes measured as decibels relative to a half-wave dipole, which means that the antenna in question is being compared to a dipole, and has a certain amount of gain relative to a dipole antenna tuned to the same operating frequency. In this case, one says the antenna has a gain of "x dBd". More often, gains are expressed relative to an isotropic radiator, which is an imaginary antenna that radiates equally in all directions. In this case one uses dBi instead of dBd. As it is impossible to build an isotropic radiator, gain measurements expressed relative to a dipole are more practical when a reference dipole aerial is used for experimental measurements. 0 dBd is often considered equal to 2.15 dBi.

Adcock Antenna

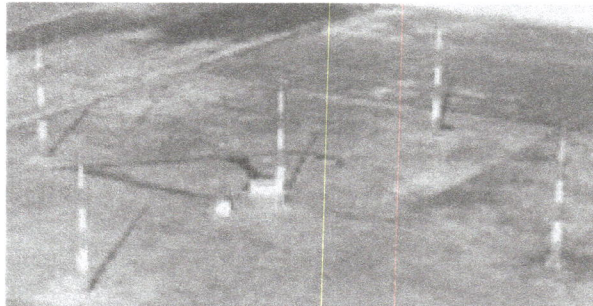

Adcock radio range ground station. Hundreds of these stations were deployed around the U.S. alone.

The Adcock antenna is an antenna array consisting of four equidistant vertical elements which can be used to transmit or receive directional radio waves.

The Adcock array was invented and patented by British engineer Frank Adcock in 1919 as British Patent No. 130,490, and has been used for a variety of applications, both civilian and military, ever since. Although originally conceived for receiving Low Frequency (LF) waves, it has also been used for transmitting, and has since been adapted for use at much higher frequencies, up to Ultra High Frequency (UHF).

In the early 1930s, the Adcock antenna (transmitting in the LF/MF bands) became a key feature of the newly created radio navigation system for aviation. The Low Frequency radio range (LFR) network, which consisted of hundreds of Adcock antenna arrays, defined the airways used by aircraft for instrument flying. The LFR remained as the main aerial navigation technology until it was replaced by the VOR system in the 1950s and 1960s.

The Adcock antenna array has been widely used commercially, and implemented in vertical antenna heights ranging from over 130 feet (40 meters) in the LFR network, to as small as 5 inches (13 cm) in tactical direction finding applications (receiving in the UHF band).

Radio Direction Finding

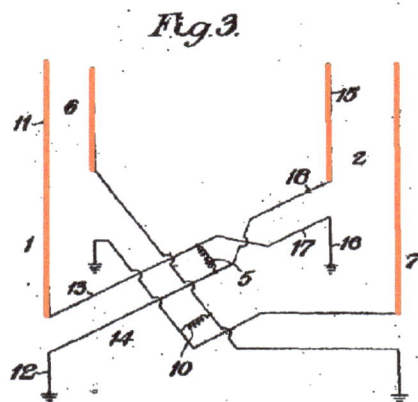

Diagram from Adcock's 1919 patent, depicting a four-element monopole antenna array; active antenna segments are marked in red.

90-foot (27 m) diagonal spacing Japanese Adcock direction finder installation for 2MHz in Rabaul

Frank Adcock originally used the antenna as a receiving antenna, to find the azimuthal direction a radio signal was coming from in order to find the location of the radio transmitter; a process called radio direction finding.

Prior to Adcock's invention, engineers had been using loop antennas to achieve directional sensitivity. They discovered that due to atmospheric disturbances and reflections, the detected signals included significant components of electromagnetic interference and distortions: horizontally polarized radiation contaminating the signal of interest and reducing the accuracy of the measurement.

Adcock—who was serving as an Army officer in the British Expeditionary Force in wartime France at the time he filed his invention—solved this problem by replacing the loop antennas with symmetrically inter-connected pairs of vertical monopole or dipole antennas of equal length. This created the equivalent of square loops, but without their horizontal members, thus eliminating sensitivity to much of the horizontally polarized distortion. The same principles remain valid today, and the Adcock antenna array and its variants are still used for radio direction finding.

Low Frequency Radio Range

In the late 1920s, the Adcock antenna was adopted for aerial navigation, in what became known as the Low Frequency radio range (LFR), or the "Adcock radio range". Hundreds of transmitting stations, each consisting of four or five Adcock antenna towers, were constructed around the U.S. and elsewhere.

The result was a network of electronic airways, which allowed pilots to navigate at night and in poor visibility, under virtually all weather conditions. The LFR remained as the main aerial navigation system in the U.S. and other countries until the 1950s, when it was replaced by VHF-based VOR technology. By the 1980s all LFR stations were decommissioned.

Loop Antenna

A shortwave loop antenna

A loop antenna is a radio antenna consisting of a loop or coil of wire, tubing, or other electrical conductor with its ends connected to a balanced transmission line (or possibly a balun). There are two distinct antenna designs: the small loop (or *magnetic loop*) with a size much smaller than a wavelength, and the much larger resonant loop antenna with a circumference close to the intended wavelength of operation.

Small loops have low radiation resistance and thus poor efficiency and are mainly used as receiving antennas at low frequencies. To increase the magnetic field in the loop and thus the efficiency, the coil of wire is often wound around a ferrite rod magnetic core; this is called a *ferrite loop* antenna. The ferrite loop is the antenna used in many AM broadcast receivers, with the exception of external loops used with AV Amplifier-Receivers and car radios; the antenna is often contained inside the radio's case. These antennas are also used for radio direction finding. In amateur radio, loop antennas are often used for low profile operating where larger antennas would be inconvenient, unsightly, or banned. Loop antennas are relatively easy to build.

A small loop antenna, also known as a magnetic loop, generally has a circumference of less than one tenth of a wavelength, in which case there will be a relatively constant current distribution along the conductor. As the frequency or the size is increased, a standing wave starts to develop in the current, and the antenna starts to acquire some of the characteristics of a resonant loop (but isn't resonant); these intermediate cases thus cannot be analyzed using the concepts developed for the small and resonant loop antennas described below.

Full-wave loop antennas are relatively large in order to be "naturally" self-resonant, with their size determined by the longest intended wavelength of operation. For reasons of size, they are typically used at higher frequencies, especially VHF and UHF, where their dimensions are more convenient. They can be viewed as a folded dipole whose parallel wires have been split, and then have been deformed into an oval or rectangle, or equivalently as two bent, half-wave dipoles connected end-to-end. In either case, full-size loops have similar characteristics to double-dipoles and folded dipoles, such as a high radiation efficiency.

Types

Full-size Loops

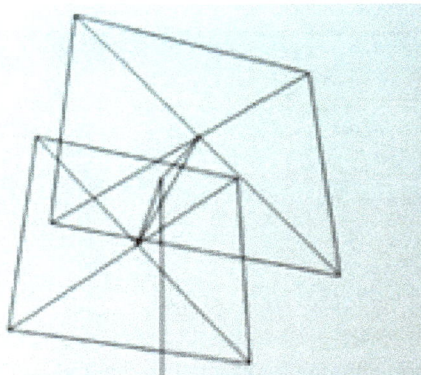

The "Quad antenna" is a resonant loop in a square shape; this one also includes a parasitic element

Although a self-resonant loop may be in the shape of a circle, distorting it into a somewhat different closed shape does not greatly alter its characteristics. For instance, the quad antenna popular in amateur radio consists of a resonant loop (and usually additional parasitic elements) in a rectangular shape, so that it can be constructed of wire strung across a supporting 'X' frame. Or a large loop can be completely collapsed into a line, in which case it is termed a folded dipole.

In the case of large loops, such as the quad or a folded dipole, the antenna's resonant frequency is determined by the wavelength that matches the circumference of the loop.

Small Loops

On the other hand, a small loop antenna is used for wavelengths much bigger than the loop itself; its radiation resistance and efficiency are instead dependent on the area enclosed by the loop (and number of turns). For a given loop area, the length of the conductor (and thus its net loss resistance) is minimized if the shape is a circle, making a circle the optimum shape for small loops.

Halo Antennas

Although it has a superficially similar appearance, the so-called halo antenna is not technically a loop since it possesses a break in the conductor opposite the feed point. Its characteristics are unlike that of either sort of loop antenna described in this article.

RF ID Pickup Loops are not Antennas

Also outside the scope of this article is the use of coupling coils for inductive (magnetic) transmission systems including LF and HF (rather than UHF) RFID tags and readers.

Although these do use radio frequencies, and involve the use of small loops (loosely described as "antennas" in the trade) which may be physically indistinguishable from the small loop antennas discussed here, such systems are not designed to transmit radio waves (electromagnetic waves). They are near field systems involving alternating magnetic fields only, and may be analyzed as poorly coupled transformer windings; their performance criteria are dissimilar to radio antennas as discussed here.

Self-resonant Loop Antennas

A loop antenna for amateur radio under construction

The large or resonant loop antenna can be seen as a folded dipole which has been reformed into a circle (or square, etc.). In order to be resonant (to have a purely resistive feed-point impedance) the loop requires a circumference approximately equal to one full wavelength (however it will also be resonant at odd multiples of a wavelength).

Contrary to the small loop antenna, the radiation is maximum in the direction *normal* to the plane of the loop (thus in two opposite directions). Therefore these loops are usually installed with the axis of the loop in the horizontal direction, and may be rotatable. Compared to a dipole or folded dipole, it transmits slightly less toward the sky or ground, giving it about a 10% higher gain in the horizontal direction. Further gain can be obtained by using a loop whose circumference is not one but 3 or 5 wavelengths. However increased gain (and a uni-directional radiation pattern) is usually obtained with an array of such elements either as a driven endfire array or in a Yagi configuration (with all but one loop being parasitic elements). The latter is widely used in amateur radio where it is referred to as a quad antenna, with the loops being square as they are usually constructed with wires held taut in between the rigid "X" structures.

The polarization of such an antenna is not obvious by looking at the loop itself, but depends on the feed point (where the transmission line is connected). If a vertically oriented loop is fed at the bottom it will be horizontally polarized; feeding it from the side will make it vertically polarized.

Small Loops

Small loop antennas are much less than a wavelength in size, and are mainly (but not always) used as receiving antennas at lower frequencies.

Small loop antenna used for receiving, consisting of about 10 turns around a 12 cm × 10 cm rectangle.

Magnetic vs. Electrical Antennas

Although a full 2.7 meters in diameter, this receiving antenna is a "small" loop for LF/MF wavelengths.

The small loop antenna is also known as a *magnetic loop* since it behaves electrically as a coil (inductor). It has a small radiation resistance due to its small size compared to the wavelength of the waves, making it an inefficient transmitting antenna. It couples to the magnetic field of the radio wave in the region near the antenna, in contrast to monopole and dipole antennas which couple to the electric field of the wave. In a receiving antenna (the main application of small loops) the oscillating magnetic field of the incoming radio wave induces a current in the wire winding by Faraday's law of induction.

Radiation Pattern and Polarization

Surprisingly, the radiation and receiving pattern of a small loop is quite opposite that of a large loop (whose circumference is close to one wavelength). Since the loop is much smaller than a wavelength, the current at any one moment is nearly constant round the circumference. By symmetry it can be seen that the voltages induced along the flat sides of the loop will cancel each other when a signal arrives along the loop axis. Therefore, there is a null in that direction. Instead, the radiation pattern peaks in directions lying in the plane of the loop, because signals received from sources in that plane do not quite cancel owing to the phase difference between the arrival of the wave at the near side and far side of the loop. Increasing that phase difference by increasing the size of the loop has a large impact in increasing the radiation resistance and the resulting antenna efficiency.

Another way of looking at a small loop as an antenna is to consider it simply as an inductive coil coupling to the magnetic field in the direction *perpendicular* to plane of the coil, according to Ampère's law. Then consider a propagating radio wave also perpendicular to that plane. Since the magnetic (and electric) fields of an electromagnetic wave in free space are transverse (no component in the direction of propagation), it can be seen that this magnetic field and that of a small loop antenna will be at right angles, and thus not coupled. For the same reason, an electromagnetic wave propagating within the plane of the loop, with its magnetic field perpendicular to that plane, *is* coupled to the magnetic field of the coil. Since the transverse magnetic and electric fields of a propagating electromagnetic wave are at right angles, the electric field of such a wave is also in the plane of the loop, and thus the antenna's *polarization* (which is always specified as being the orientation of the electric, not the magnetic field) is said to be in that plane.

Thus mounting the loop in a horizontal plane will produce an omnidirectional antenna which is horizontally polarized; mounting the loop vertically yields a weakly directional antenna with vertical polarization and sharp nulls along the axis of the loop.

Small Loop Receiving Antennas

AM broadcast radios (and other consumer low frequency receivers) typically use small loop antennas; a variable capacitor connected across the loop forms a tuned circuit that also tunes the receivers input stage as that capacitor tracks the main tuning. A multiband receiver may contain tap points along the loop winding in order to tune the loop antenna at widely different frequencies. In older AM radios, the antenna might consist of dozens of turns of wire mounted on the back wall of the radio (a *frame antenna*).

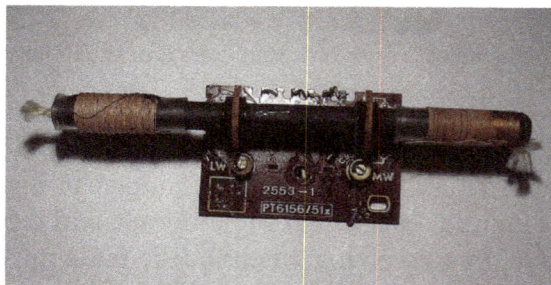

Ferrite loopstick antenna from an AM radio having two windings, one for long wave and one for medium wave (AM broadcast) reception. About 10 cm long. Ferrite antennas are usually enclosed inside the radio receiver.

In modern radios, a ferrite loop antenna is used, consisting of fine wire wound on a ferrite rod. Litz wire is often used to reduce skin effect losses. The ferrite rod increases the magnetic permeability, allowing the physically small antenna to have a larger effective area. Other names for this type of antenna are *loopstick antenna, ferrite rod antenna, ferrite rod aerial, Ferroceptor,* or *ferrod antenna*.

Small loop antennas are lossy and inefficient, but they can make practical receiving antennas in the medium-wave (520–1610 kHz) band and below, where the antenna inefficiency is masked by large amounts of atmospheric noise. Loop antennas are often wound with litz wire to reduce skin effect losses.

Antenna Efficiency

Since a small loop antenna is essentially a coil, its electrical impedance is inductive, with an inductive reactance much greater than its radiation resistance. In order to couple to a transmitter or receiver, the inductive reactance is normally canceled with a parallel capacitance. Since a good loop antenna will have a high Q factor, this capacitor must be variable and is adjusted along with the receiver's tuning.

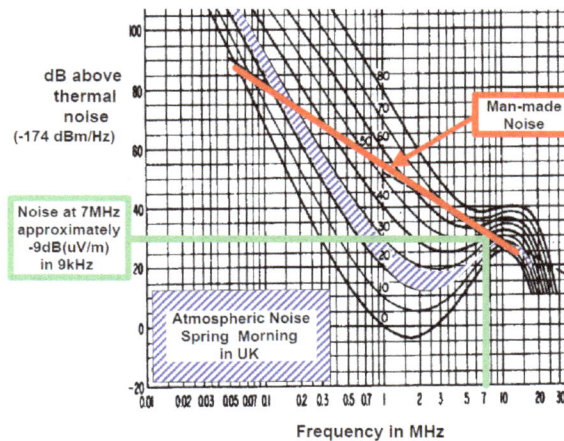

Amount of atmospheric noise for LF, MF, and HF spectrum according CCIR 322

The radiation resistance R_R of a small loop is generally much smaller than the loss resistance R_L due to the conductors comprising the loop, leading to a poor antenna efficiency. Consequently, most of the transmitted or received power will be dissipated as heat.

So much wasted signal power is a disaster for a transmitting antenna, however in a receiving antenna the inefficiency is not important at frequencies below about 10 MHz . At those lower frequencies atmospheric noise (static) and man-made noise (interference) dominate over the noise generated inside the receiver itself (*thermal* or *Johnson noise*). Any increase in signal strength increases both the signal and the external noise in equal proportion, leaving the signal-to-noise ratio unchanged. (CCIR 258; CCIR 322.)

For example, at 1 MHz the man-made noise might be 55 dB above the thermal noise floor. If a small loop antenna's loss is 50 dB (as if the antenna included a 50 dB attenuator) the electrical inefficiency of that antenna will have little influence on the receiving system's signal-to-noise ratio.

In contrast, at quieter frequencies above about 20 MHz an antenna with a 50 dB loss could degrade the received signal-to-noise ratio by up to 50 dB, resulting in terrible performance. Copper losses are often minimized by the use of spiderweb or basket winding construction and Litz wire.

Insensitivity to Locally Generated Interference

Due to its direct coupling to the magnetic field, unlike most other antenna types, the small loop is relatively insensitive to electric-field noise from nearby sources. No matter how close the electrical interference is to the loop, its effect will not be much greater than if it were a quarter wavelength away. This is valuable since most sources of interference with radio frequency content, such as sparking at commutators or corona discharge, start as electric fields; the resulting field only develops equally strong electric and magnetic parts a quarter-wavelength from the source. It is also convenient that at the long wavelengths where small loops excel, the quarter-wavelength region is larger than the size of a large house, reducing issues with interference generated in the home.

Small loops are especially used in the AM broadcast band and generally at lower frequencies where resonant antennas are of an impractical size. At those frequencies the near-field is physically quite large. This provides a considerable advantage for loop antennas which are relatively insensitive to the main interference sources encountered.

The same principle makes a small loop particularly sensitive to sources of *magnetic* noise in its near field. Likewise, a Hertzian (short) dipole couples directly with the electric field and is relatively immune to locally produced magnetic noise. However at radio frequencies nearby sources of magnetic interference are generally not an issue. In either case the small antenna's immunity does not extend to noise sources outside of the near field: Noise sources over one wavelength distant, whether originating as an electric or magnetic field, are received simply as electromagnetic waves. Noise from outside any antenna's near field will be received equally well by any antenna sensitive to a radio transmitter from the direction of that noise source.

Receiver Input Tuning

Small loop receiving antennas are also almost always resonated using a parallel plate capacitor, which makes their reception narrow-band, sensitive only to a very specific frequency. This allows the antenna, in conjunction with a (variable) tuning capacitor, to act as a tuned input stage to the receiver's front-end, in lieu of a coil.

Small Loops as Transmitting Antennas

Due to their small radiation resistance and consequent electrical inefficiency, small loops are seldom used as transmitting antennas, where one is trying to couple most of the transmitter's power to the electromagnetic field. Nevertheless, small loops are sometimes used in applications in which a self-resonant antenna (with elements around a quarter of a wavelength in size) would simply be too large to be practical.

Since *any* antenna much smaller than a wavelength suffers from inefficiency, a loop might not be the worst choice for medium wave and lower frequencies. Small loops (typically 18 to 39 inches in diameter when used from 29.7–7.0 MHz) are becoming popular as transmitting antennas, as well as receiving antennas.

As for any antenna system, efficient electrical coupling requires impedance matching. A small loop tuned with a parallel capacitor results in a large impedance compared to the impedance of a standard transmission line or transmitter, and a small loop tuned with a series capacitor results in a small impedance. In addition to other common impedance matching techniques, transmitting loops are sometimes impedance matched by connecting the feedline to a smaller *feed loop* inside the area surrounded by the main loop. Typical feed loops are $1/_8$ to $1/_5$ the size of the antenna's main loop. The combination is in effect a step-up transformer, with power in the near-field inductively coupled from the feed loop to the main loop, which itself is connected to the resonating capacitor and is responsible for radiating most of the power.

Radiation Efficiency

The radiation efficiency of a loop antenna can be greatly improved by making the main loop large – as compared, say, to a loop only used for receiving. Ideally, the radiating loop's circumference should be greater than $\frac{1}{10}$ of a wavelength.

The increased size of the now not-so-small loop alters its radiation pattern, as the assumption of currents being totally in phase along the circumference of the loop begins to break down. In addition to making the geometric loop larger, efficiency is also increased by using larger conductors in order to reduce the loss resistance, and plating the outer conductor surfaces with silver or non-anodized aluminum.

Use for Land-mobile Radio

Small loops are used in land-mobile radio (mostly military) at frequencies between 3–7 MHz, because of their ability to direct energy upwards, unlike a conventional whip antenna. This enables Near Vertical Incidence Skywave (NVIS) communication up to 300 km in mountainous regions. In this case a typical radiation efficiency of around 1% is acceptable because signal paths can be established with 1 Watt of radiated power or less when a transmitter generating 100 Watts is used. In military use, the antenna elements can be 2–3 inches in diameter.

Power Limits

One practical issue with small loops as transmitting antennas is that the loop not only has a very large current going through it, but also has a very high voltage on its terminals, typically kilo-Volts when fed with only a few Watts of transmitter power. This requires a rather expensive and physically large resonating capacitor with a large breakdown voltage, in addition to having minimal dielectric loss (normally requiring an air-gap capacitor).

To keep a balanced perspective, it is important to note that a vertical or dipole antenna that is short compared to a wavelength, matched using a small loading coil *also* has a high voltage present at the loading coil. The difference being that since the loop antenna is already physically large in order to reduce loss and carry high current, high voltage breakdown is not usually as much of an issue.

Direction Finding with Loops

Loop antenna, receiver, and accessories used in amateur radio direction finding at 80 meter wavelength (3.5 MHz).

Since the directional response of small loop antennas includes a sharp null in the direction normal to the plane of the loop, they are used in radio direction finding at longer wavelengths.

The procedure is to rotate the loop antenna to find the direction where the signal vanishes – the "null" direction. Since the null occurs at two opposite directions along the axis of the loop, other means must be employed to determine which side of the antenna the "nulled" signal is on. One method is to rely on a second loop antenna located at a second location, or to move the receiver to that other location, thus relying on triangulation.

Instead of triangulation, a second dipole or vertical antenna can be electrically combined with a loop or a loopstick antenna. Called a *sense antenna*, connecting the second antenna changes the combined radiation pattern to a cardioid, with a null in only one, less precise direction. The general direction of the transmitter can be determined using the sense antenna, and then disconnecting the sense antenna returns the sharp nulls in the loop antenna pattern, allowing a precise bearing to be determined.

Cage Aerial

A cage antenna (british cage aerial) is a radio antenna that consists of the top portion of a tower or mast and of several parallel wires, which are radially arranged around the lower part of the mast. One advantage of the cage aerial is that the supporting tower can be grounded, allowing it to be used for other radio services, such as a support for VHF or UHF antennas. A grounded tower also simplifies the installation of aircraft warning lamps. Cage aerials have been built in different variants for broadcasting stations in the longwave and mediumwave bands.

The cage is electrically one-quarter of the operating wavelength. It is connected to the mast at its upper end. That way it isolates the lower part of the mast ($\lambda/4$ stub) and makes the upper part of the mast the radiator. Very often the typical height of such an antenna is no problem as the height of the mast is selected for the TV or FM antennas on top.

Example: At 1000 kHz the wavelength is 300 m. Therefore the minimum length of the cage antenna is a bit more than 150 m; 75 m for the radiator, 75 m for the cage and a few metres to make the lower end of the cage inaccessible from the ground, as the lower end of the cage carries a very high RF voltage. This type of antenna is known in America as a "folded unipole", which has been extensively studied by John H. Mullaney.

Parabolic Antenna

A large parabolic satellite communications antenna at Erdfunkstelle Raisting, the biggest facility for satellite communication in the world, in Raisting, Bavaria, Germany. It has a Cassegrain type feed.

A parabolic antenna is an antenna that uses a parabolic reflector, a curved surface with the cross-sectional shape of a parabola, to direct the radio waves. The most common form is shaped like a dish and is popularly called a dish antenna or parabolic dish. The main advantage of a parabolic antenna is that it has high directivity. It functions similarly to a searchlight or flashlight reflector to direct the radio waves in a narrow beam, or receive radio waves from one particular direction only. Parabolic antennas have some of the highest gains, meaning that they can produce the narrowest beamwidths, of any antenna type. In order to achieve narrow beamwidths, the parabolic reflector must be much larger than the wavelength of the radio waves used, so parabolic antennas are used in the high frequency part of the radio spectrum, at UHF and microwave (SHF) frequencies, at which the wavelengths are small enough that conveniently-sized reflectors can be used.

Parabolic antennas are used as high-gain antennas for point-to-point communications, in applications such as microwave relay links that carry telephone and television signals between nearby cities, wireless WAN/LAN links for data communications, satellite communications and spacecraft communication antennas. They are also used in radio telescopes.

The other large use of parabolic antennas is for radar antennas, in which there is a need to transmit a narrow beam of radio waves to locate objects like ships, airplanes, and guided missiles. With the advent of home satellite television receivers, parabolic antennas have become a common feature of the landscapes of modern countries.

The parabolic antenna was invented by German physicist Heinrich Hertz during his discovery of radio waves in 1887. He used cylindrical parabolic reflectors with spark-excited dipole antennas at their focus for both transmitting and receiving during his historic experiments.

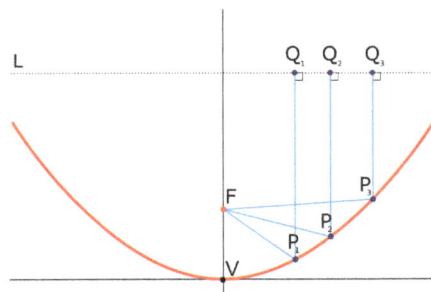

Parabolic antennas are based on the geometrical property of the paraboloid that the paths FP_1Q_1, FP_2Q_2, FP_3Q_3 are all the same length. So a spherical wavefront emitted by a feed antenna at the dish's focus F will be reflected into an outgoing plane wave L travelling parallel to the dish's axis VF.

Design

The operating principle of a parabolic antenna is that a point source of radio waves at the focal point in front of a paraboloidal reflector of conductive material will be reflected into a collimated plane wave beam along the axis of the reflector. Conversely, an incoming plane wave parallel to the axis will be focused to a point at the focal point.

A typical parabolic antenna consists of a metal parabolic reflector with a small feed antenna suspended in front of the reflector at its focus, pointed back toward the reflector. The reflector is a metallic surface formed into a paraboloid of revolution and usually truncated in a circular rim that forms the diameter of the antenna. In a transmitting antenna, radio frequency current from a transmitter is supplied through a transmission line cable to the feed antenna, which converts it into radio waves. The radio waves are emitted back toward the dish by the feed antenna and reflect off the dish into a parallel beam. In a receiving antenna the incoming radio waves bounce off the dish and are focused to a point at the feed antenna, which converts them to electric currents which travel through a transmission line to the radio receiver.

Parabolic Reflector

Wire grid-type parabolic antenna used for MMDS data link at a frequency of 2.5-2.7 GHz. It is fed by a vertical dipole under the small aluminum reflector on the boom. It radiates vertically polarized microwaves.

The reflector can be of sheet metal, metal screen, or wire grill construction, and it can be either a circular "dish" or various other shapes to create different beam shapes. A metal screen reflects radio waves as well as a solid metal surface as long as the holes are smaller than one-tenth of a wavelength, so screen reflectors are often used to reduce weight and wind loads on the dish. To achieve the maximum gain, it is necessary that the shape of the dish be accurate within a small fraction of a wavelength, to ensure the waves from different parts of the antenna arrive at the focus in phase. Large dishes often require a supporting truss structure behind them to provide the required stiffness.

A reflector made of a grill of parallel wires or bars oriented in one direction acts as a *polarizing filter* as well as a reflector. It only reflects linearly polarized radio waves, with the electric field parallel to the grill elements. This type is often used in radar antennas. Combined with a linearly polarized feed horn, it helps filter out noise in the receiver and reduces false returns.

Since a shiny metal parabolic reflector can also focus the sun's rays, and most dishes could concentrate enough solar energy on the feed structure to severely overheat it if they happened to be pointed at the sun, solid reflectors are always given a coat of flat paint.

Feed Antenna

The feed antenna at the reflector's focus is typically a low-gain type such as a half-wave dipole or more often a small horn antenna called a feed horn. In more complex designs, such as the Cassegrain and Gregorian, a secondary reflector is used to direct the energy into the parabolic reflector from a feed antenna located away from the primary focal point. The feed antenna is connected to the associated radio-frequency (RF) transmitting or receiving equipment by means of a coaxial cable transmission line or waveguide.

At the microwave frequencies used in many parabolic antennas, waveguide is required to conduct the microwaves between the feed antenna and transmitter or receiver. Because of the high cost of wave-guide runs, in many parabolic antennas the RF front end electronics of the receiver is located at the feed antenna, and the received signal is converted to a lower intermediate frequency (IF) so it can be conducted to the receiver through cheaper coaxial cable. This is generally called a low noise amplifier (LNA). Similarly, in transmitting dishes, the microwave transmitter may be located at the feed point.

An advantage of parabolic antennas is that most of the structure of the antenna (all of it except the feed antenna) is nonresonant, so it can function over a wide range of frequencies, that is a wide bandwidth. All that is necessary to change the frequency of operation is to replace the feed antenna with one that works at the new frequency. Some parabolic antennas transmit or receive at multiple frequencies by having several feed antennas mounted at the focal point, close together.

Dish Parabolic Antennas

Shrouded microwave relay dishes on a communications tower in Australia.

Offset Gregorian antenna used in the Allen Telescope Array, a radio telescope at the University of California at Berkeley, USA.

A satellite television dish, an example of an offset fed dish.

Cassegrain satellite communication antenna in Sweden.

Shaped-beam Parabolic Antennas

Vertical "orange peel" antenna for military height finder radar, Germany.

Early cylindrical parabolic antenna, 1931, Nauen, Germany.

Air traffic control radar antenna, near Hannover, Germany.

"Orange peel" antenna for air search radar, Finland.

Types

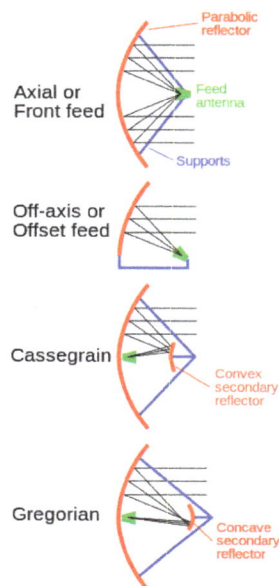

Main types of parabolic antenna feeds.

Parabolic antennas are distinguished by their shapes:

- *Paraboloidal or dish* – The reflector is shaped like a paraboloid truncated in a circular rim. This is the most common type. It radiates a narrow pencil-shaped beam along the axis of the dish.

  - *Shrouded dish* – Sometimes a cylindrical metal shield is attached to the rim of the dish. The shroud shields the antenna from radiation from angles outside the main beam axis, reducing the sidelobes. It is sometimes used to prevent interference in terrestrial microwave links, where several antennas using the same frequency are located close together. The shroud is coated inside with microwave absorbent material. Shrouds can reduce back lobe radiation by 10 dB.

- *Cylindrical* – The reflector is curved in only one direction and flat in the other. The radio waves come to a focus not at a point but along a line. The feed is sometimes a dipole antenna located along the focal line. Cylindrical parabolic antennas radiate a fan-shaped beam, narrow in the curved dimension, and wide in the uncurved dimension. The curved ends of the reflector are sometimes capped by flat plates, to prevent radiation out the ends, and this is called a *pillbox* antenna.

- *Shaped-beam antennas* – Modern reflector antennas can be designed to produce a beam or beams of a particular shape, rather than just the narrow "pencil" or "fan" beams of the simple dish and cylindrical antennas above. Two techniques are used, often in combination, to control the shape of the beam:

  o *Shaped reflectors* – The parabolic reflector can be given a noncircular shape, and/ or different curvatures in the horizontal and vertical directions, to alter the shape of the beam. This is often used in radar antennas. As a general principle, the wider the antenna is in a given transverse direction, the narrower the radiation pattern will be in that direction.

    ▪ *"Orange peel" antenna* – Used in search radars, this is a long narrow antenna shaped like the letter "C". It radiates a narrow vertical fan shaped beam.

Array of multiple feed horns on a German airport surveillance radar antenna to control the elevation angle of the beam

  o *Arrays of feeds* – In order to produce an arbitrary shaped beam, instead of one feed horn, an array of feed horns clustered around the focal point can be used. Array-fed antennas are often used on communication satellites, particularly direct broadcast satellites, to create a downlink radiation pattern to cover a particular continent or coverage area. They are often used with secondary reflector antennas such as the Cassegrain.

Parabolic antennas are also classified by the type of *feed*, that is, how the radio waves are supplied to the antenna:

- *Axial or front feed* – This is the most common type of feed, with the feed antenna located in front of the dish at the focus, on the beam axis, pointed back toward the dish. A disadvantage of this type is that the feed and its supports block some of the beam, which limits the aperture efficiency to only 55–60%.

- *Off-axis or offset feed* – The reflector is an asymmetrical segment of a paraboloid, so the focus, and the feed antenna, are located to one side of the dish. The purpose of this design is to move the feed structure out of the beam path, so it does not block the beam. It is widely used in home satellite television dishes, which are small enough that the feed structure would otherwise block a significant percentage of the signal. Offset feed can also be used in multiple reflector designs such as the Cassegrain and Gregorian, below.

- *Cassegrain* – In a Cassegrain antenna, the feed is located on or behind the dish, and radiates forward, illuminating a convex hyperboloidal secondary reflector at the focus of the dish. The radio waves from the feed reflect back off the secondary reflector to the dish, which forms the outgoing beam. An advantage of this configuration is that the feed, with its waveguides and "front end" electronics does not have to be suspended in front of the dish, so it is used for antennas with complicated or bulky feeds, such as large satellite communication antennas and radio telescopes. Aperture efficiency is on the order of 65–70%.

- *Gregorian* – Similar to the Cassegrain design except that the secondary reflector is concave, (ellipsoidal) in shape. Aperture efficiency over 70% can be achieved.

Feed pattern

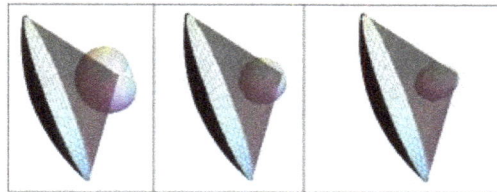

Effect of the feed antenna radiation pattern *(small pumpkin-shaped surface)* on spillover. *Left:* With a low gain feed antenna, significant parts of its radiation fall outside the dish. *Right:* With a higher gain feed, almost all its radiation is emitted within the angle of the dish.

The radiation pattern of the feed antenna has to be tailored to the shape of the dish, because it has a strong influence on the *aperture efficiency*, which determines the antenna gain. Radiation from the feed that falls outside the edge of the dish is called *"spillover"* and is wasted, reducing the gain and increasing the backlobes, possibly causing interference or (in receiving antennas) increasing susceptibility to ground noise. However, maximum gain is only achieved when the dish is uniformly "illuminated" with a constant field strength to its edges. So the ideal radiation pattern of a feed antenna would be a constant field strength throughout the solid angle of the dish, dropping abruptly to zero at the edges. However, practical feed antennas have radiation patterns that drop off gradually at the edges, so the feed antenna is a compromise between acceptably low spillover and adequate illumination. For most front feed horns, optimum illumination is achieved when the power radiated by the feed horn is 10 dB less at the dish edge than its maximum value at the center of the dish.

Polarization

The pattern of electric and magnetic fields at the mouth of a parabolic antenna is simply a scaled up image of the fields radiated by the feed antenna, so the polarization is determined by the feed antenna. In order to achieve maximum gain, the feed antenna in the transmitting and receiving antenna must have the same polarization. For example, a vertical dipole feed antenna will radiate a beam of radio waves with their electric field vertical, called vertical polarization. The receiving feed antenna must also have vertical polarization to receive them; if the feed is horizontal (horizontal polarization) the antenna will suffer a severe loss of gain.

To increase the data rate, some parabolic antennas transmit two separate radio channels on the same frequency with orthogonal polarizations, using separate feed antennas; this is called a *dual polarization* antenna. For example, satellite television signals are transmitted from the satellite on two separate channels at the same frequency using right and left circular polarization. In a home satellite dish, these are received by two small monopole antennas in the feed horn, oriented at right angles. Each antenna is connected to a separate receiver.

If the signal from one polarization channel is received by the oppositely polarized antenna, it will cause crosstalk that degrades the signal-to-noise ratio. The ability of an antenna to keep these orthogonal channels separate is measured by a parameter called *cross polarization discrimination* (XPD). In a transmitting antenna, XPD is the fraction of power from an antenna of one polarization radiated in the other polarization. For example, due to minor imperfections a dish with a vertically polarized feed antenna will radiate a small amount of its power in horizontal polarization; this fraction is the XPD. In a receiving antenna, the XPD is the ratio of signal power received of the opposite polarization to power received in the same antenna of the correct polarization, when the antenna is illuminated by two orthogonally polarized radio waves of equal power. If the antenna system has inadequate XPD, cross polarization interference cancelling (XPIC) digital signal processing algorithms can often be used to decrease crosstalk.

Dual Reflector Shaping

In the Cassegrain and Gregorian antennas, the presence of two reflecting surfaces in the signal path offers additional possibilities for improving performance. When the highest performance is required, a technique called "dual reflector shaping" may be used. This involves changing the shape of the sub-reflector to direct more signal power to outer areas of the dish, to map the known pattern of the feed into a uniform illumination of the primary, to maximize the gain. However, this results in a secondary that is no longer precisely hyperbolic (though it is still very close), so the constant phase property is lost. This phase error, however, can be compensated for by slightly tweaking the shape of the primary mirror. The result is a higher gain, or gain/spillover ratio, at the cost of surfaces that are trickier to fabricate and test. Other dish illumination patterns can also be synthesized, such as patterns with high taper at the dish edge for ultra-low spillover sidelobes, and patterns with a central "hole" to reduce feed shadowing.

History

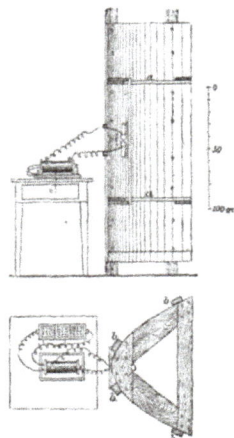

The first parabolic antenna, built by Heinrich Hertz in 1888.

The idea of using parabolic reflectors for radio antennas was taken from optics, where the power of a parabolic mirror to focus light into a beam has been known since classical antiquity. The designs of some specific types of parabolic antenna, such as the Cassegrain and Gregorian, come from similarly named analogous types of reflecting telescope, which were invented by astronomers during the 15th century.

German physicist Heinrich Hertz constructed the world's first parabolic reflector antenna in 1888. The antenna was a cylindrical parabolic reflector made of zinc sheet metal supported by a wooden frame, and had a spark-gap excited dipole as a feed antenna along the focal line. Its aperture was 2 meters high by 1.2 meters wide, with a focal length of 0.12 meters, and was used at an operating frequency of about 450 MHz. With two such antennas, one used for transmitting and the other for receiving, Hertz demonstrated the existence of radio waves which had been predicted by James Clerk Maxwell some 22 years earlier. However, the early development of radio was limited to lower frequencies at which parabolic antennas were unsuitable, and they were not widely used until after World War 2, when microwave frequencies began to be exploited.

Italian radio pioneer Guglielmo Marconi used a parabolic reflector during the 1930s in investigations of UHF transmission from his boat in the Mediterranean. In 1931 a 1.7 GHz microwave relay telephone link across the English Channel using 10 ft. (3 meter) diameter dishes was demonstrated. The first large parabolic antenna, a 9 m dish, was built in 1937 by pioneering radio astronomer Grote Reber in his backyard, and the sky survey he did with it was one of the events that founded the field of radio astronomy.

The development of radar during World War II provided a great impetus to parabolic antenna research, and saw the evolution of shaped-beam antennas, in which the curve of the reflector is different in the vertical and horizontal directions, tailored to produce a beam with a particular shape. After the war very large parabolic dishes were built as radio telescopes. The 100 meter Green Bank Radio Telescope at Green Bank, West Virginia, the first version of which was completed in 1962, is currently the world's largest fully steerable parabolic dish.

During the 1960s dish antennas became widely used in terrestrial microwave relay communication networks, which carried telephone calls and television programs across continents. The first parabolic antenna used for satellite communications was constructed in 1962 at Goonhilly in Cornwall, England to communicate with the Telstar satellite. The Cassegrain antenna was developed in Japan in 1963 by NTT, KDDI and Mitsubishi Electric. The advent in the 1970s of computer design tools such as NEC capable of calculating the radiation pattern of parabolic antennas has led to the development of sophisticated asymmetric, multireflector and multifeed designs in recent years.

Gain

The directive qualities of an antenna are measured by a dimensionless parameter called its gain, which is the ratio of the power received by the antenna from a source along its beam axis to the power received by a hypothetical isotropic antenna. The gain of a parabolic antenna is:

$$G = \frac{4\pi A}{\lambda^2} e_A = \left(\frac{\pi d}{\lambda}\right)^2 e_A$$

The second-largest "dish" antenna in the world, the radio telescope at Arecibo Observatory, Puerto Rico, 305 meters in diameter. It has a gain of about 10 million, or 70 dBi, at 2.38 GHz. The dish is constructed into a valley in the landscape, so it is not steerable. To steer the antenna to point to different regions in the sky, the feed antenna suspended by cables over the dish is moved. The dish actually has a spherical rather than a parabolic shape, which reduces the aberrations caused by moving the feed point, but also means that the received energy comes to a focus along a line rather than a single point. This spherical aberration can be corrected by a secondary reflector of the proper shape.

where:

A is the area of the antenna aperture, that is, the mouth of the parabolic reflector. For a circular dish antenna, $A = \pi d^2 / 4$, giving the second formula above.

d is the diameter of the parabolic reflector, if it is circular

λ is the wavelength of the radio waves.

e_A is a dimensionless parameter between 0 and 1 called the *aperture efficiency*. The aperture efficiency of typical parabolic antennas is 0.55 to 0.70.

It can be seen that, as with any *aperture antenna*, the larger the aperture is, compared to the wavelength, the higher the gain. The gain increases with the square of the ratio of aperture width to wavelength, so large parabolic antennas, such as those used for spacecraft communication and radio telescopes, can have extremely high gain. Applying the above formula to the 25-meter-diameter antennas often used in radio telescope arrays and satellite ground antennas at a wavelength of 21 cm (1.42 GHz, a common radio astronomy frequency), yields an approximate maximum gain of 140,000 times or about 50 dBi (decibels above the isotropic level).

Aperture efficiency e_A is a catchall variable which accounts for various losses that reduce the gain of the antenna from the maximum that could be achieved with the given aperture. The major factors reducing the aperture efficiency in parabolic antennas are:

- *Feed spillover* - Some of the radiation from the feed antenna falls outside the edge of the dish and so doesn't contribute to the main beam.

- *Feed illumination taper* - The maximum gain for any aperture antenna is only achieved when the intensity of the radiated beam is constant across the entire aperture area. However the radiation pattern from the feed antenna usually tapers off toward the outer part of the dish, so the outer parts of the dish are "illuminated" with a lower intensity of radiation. Even if the feed provided constant illumination across the angle subtended by the dish, the outer parts of the dish are farther away from the feed antenna than the inner parts, so the intensity would drop off with distance from the center. So the intensity of the beam radiated by a parabolic antenna is maximum at the center of the dish and falls off with distance from the axis, reducing the efficiency.

- *Aperture blockage* - In front-fed parabolic dishes where the feed antenna is located in front of the dish in the beam path (and in Cassegrain and Gregorian designs as well), the feed structure and its supports block some of the beam. In small dishes such as home satellite dishes, where the size of the feed structure is comparable with the size of the dish, this can seriously reduce the antenna gain. To prevent this problem these types of antennas often use an *offset* feed, where the feed antenna is located to one side, outside the beam area. The aperture efficiency for these types of antennas can reach 0.7 to 0.8.

- *Shape errors* - random surface errors in the shape of the reflector reduce efficiency. The loss is approximated by Ruze's Equation.

For theoretical considerations of mutual interference (at frequencies between 2 and c. 30 GHz - typically in the Fixed Satellite Service) where specific antenna performance has not been defined, a *reference antenna* based on Recommendation ITU-R S.465 is used to calculate the interference, which will include the likely sidelobes for off-axis effects.

Radiation Pattern

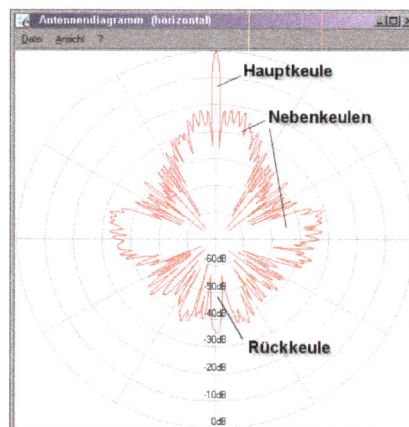

Radiation pattern of a German parabolic antenna. The main lobe *(top)* is only a few degrees wide. The sidelobes are all at least 20 dB below (1/100 the power density of) the main lobe, and most are 30 dB below.

In parabolic antennas, virtually all the power radiated is concentrated in a narrow main lobe along the antenna's axis. The residual power is radiated in sidelobes, usually much smaller, in other directions. Because in parabolic antennas the reflector aperture is much larger than the wavelength, due to diffraction there are usually many narrow sidelobes, so the sidelobe pattern is complex. There is also usually a backlobe, in the opposite direction to the main lobe, due to the spillover radiation from the feed antenna that misses the reflector.

Beamwidth

The angular width of the beam radiated by high-gain antennas is measured by the *half-power beam width* (HPBW), which is the angular separation between the points on the antenna radiation pattern at which the power drops to one-half (-3 dB) its maximum value. For parabolic antennas, the HPBW θ is given by:

$$\theta = k\lambda / d$$

where k is a factor which varies slightly depending on the shape of the reflector and the feed illumination pattern. For an ideal uniformly illuminated parabolic reflector and θ in degrees, k would be 57.3 (the number of degrees in a radian). For a "typical" parabolic antenna k is approximately 70.

For a typical 2 meter satellite dish operating on C band (4 GHz), this formula gives a beamwidth of about 2.6°. For the Arecibo antenna at 2.4 GHz the beamwidth is 0.028°. It can be seen that parabolic antennas can produce very narrow beams, and aiming them can be a problem. Some parabolic dishes are equipped with a boresight so they can be aimed accurately at the other antenna.

It can be seen there is an inverse relation between gain and beam width. By combining the beamwidth equation with the gain equation, the relation is:

$$G = \left(\frac{\pi k}{\theta}\right)^2 e_A$$

Conformal Antenna

In radio communication and avionics a conformal antenna or conformal array is a flat radio antenna which is designed to conform or follow some prescribed shape, for example a flat curving antenna which is mounted on or embedded in a curved surface. Conformal antennas were developed in the 1980s as avionics antennas integrated into the curving skin of military aircraft to reduce aerodynamic drag, replacing conventional antenna designs which project from the aircraft surface. Military aircraft and missiles are the largest application of conformal antennas, but they are also used in some civilian aircraft, military ships and land vehicles. As the cost of the required processing technology comes down, they are being considered for use in civilian applications such as train antennas, car radio antennas, and cellular base station antennas, to save space and also to make the antenna less visually intrusive by integrating it into existing objects.

How it Works

Conformal antennas are a form of phased array antenna. They are composed of an array of many identical small flat antenna elements, such as dipole, horn, or patch antennas, covering the surface. At each antenna the current from the transmitter passes through a phase shifter device which are all controlled by a microprocessor (computer). By controlling the phase of the feed current, the nondirectional radio waves emitted by the individual antennas can be made to combine in front of the antenna by the process of interference, forming a strong beam (or beams) of radio waves pointed in any desired direction. In a receiving antenna the weak individual radio signals received by each antenna element are combined in the correct phase to enhance signals coming from a particular direction, so the antenna can be made sensitive to the signal from a particular station and reject interfering signals from other directions.

In a conventional phased array the individual antenna elements are mounted on a flat surface. In a conformal antenna, they are mounted on a curved surface, and the phase shifters also compensate for the different phase shifts caused by the varying path lengths of the radio waves due to the location of the individual antennas on the curved surface. Because the individual antenna elements must be small, conformal arrays are typically limited to high frequencies in the UHF or microwave range, where the wavelength of the waves is small enough that small antennas can be used.

Dielectric Resonator Antenna

A dielectric resonator antenna (DRA) is a radio antenna mostly used at microwave frequencies and higher, that consists of a block of ceramic material of various shapes, the dielectric resonator, mounted on a metal surface, a ground plane. Radio waves are introduced into the inside of the resonator material from the transmitter circuit and bounce back and forth between the resonator walls, forming standing waves. The walls of the resonator are partially transparent to radio waves, allowing the radio power to radiate into space.

An advantage of dielectric resonator antennas is they lack metal parts, which become lossy at high frequencies, dissipating energy. So these antennas can have lower losses and be more efficient than metal antennas at high microwave and millimeter wave frequencies. Dielectric waveguide antennas are used in some compact portable wireless devices, and military millimeter-wave radar equipment. The antenna was first proposed by Robert Richtmyer in 1939. In 1982, Long et al. did the first design and test of dielectric resonator antennas considering a leaky waveguide model assuming magnetic conductor model of the dielectric surface.

An antenna like effect is achieved by periodic swing of electrons from its capacitive element to the ground plane which behaves like an inductor. The authors further argued that the operation of a dielectric antenna resembles the antenna conceived by Marconi, the only difference is that inductive element is replaced by the dielectric material.

Features

Dielectric resonator antennas offer the following attractive features:

- The dimension of a DRA is of the order of $\dfrac{\lambda_0}{\sqrt{\epsilon_r}}$, where λ_0 is the free-space wavelength and ϵ_r is the dielectric constant of the resonator material. Thus, by choosing a high value of ϵ_r ($\epsilon_r \approx 10-100$), the size of the DRA can be significantly reduced.

- There is no inherent conductor loss in dielectric resonators. This leads to high radiation efficiency of the antenna. This feature is especially attractive for millimeter (mm)-wave antennas, where the loss in metal fabricated antennas can be high.

- DRAs offer simple coupling schemes to nearly all transmission lines used at microwave and mm-wave frequencies. This makes them suitable for integration into different planar technologies. The coupling between a DRA and the planar transmission line can be easily controlled by varying the position of the DRA with respect to the line. The performance of DRA can therefore be easily optimized experimentally.

- The operating bandwidth of a DRA can be varied over a wide range by suitably choosing resonator parameters. For example, the bandwidth of the lower order modes of a DRA can be easily varied from a fraction of a percent to about 20% or more by the suitable choice of the dielectric constant of the material and/or by strategic shaping of the DRA element.

- Use of multiple modes radiating identically has also been successfully addressed.

- Each mode of a DRA has a unique internal and associated external field distribution. Therefore, different radiation characteristics can be obtained by exciting different modes of a DRA.

Antenna Array

Large planar array antenna of a Russian air defense radar, the Nebo-M. It consists of 168 Yagi folded dipole antennas driven in phase. It radiates a narrow beam of radio waves perpendicular to the antenna.

An antenna array (or array antenna) is a set of multiple connected antennas which work together as a single antenna, to transmit or receive radio waves. The individual antenna elements are connected to a single receiver or transmitter by feedlines that feed the power to the elements in a specific phase relationship. The radio waves radiated by each individual antenna combine and superpose, adding together (interfering constructively) to enhance the power radiated in desired directions, and cancelling (interfering destructively) to reduce the power radiated in other directions. Similarly, when used for receiving, the separate radio frequency currents from the individual antennas combine in the receiver with the correct phase relationship to enhance signals received from the desired directions and cancel signals from undesired directions.

An antenna array can achieve higher gain (directivity), that is a narrower beam of radio waves, than could be achieved by a single antenna. In general, the larger the number of individual antenna elements used, the higher the gain and the narrower the beam. Some antenna arrays (such as military phased array radars) are composed of thousands of individual antennas. Arrays can be used to achieve higher gain, to give path diversity (also called MIMO) which increases communication reliability, to cancel interference from specific directions, to steer the radio beam electronically to point in different directions, and for radio direction finding (RDF).

The term antenna array most commonly means a *driven array* consisting of multiple identical driven elements all connected to the receiver or transmitter, often half-wave dipoles fed in phase.

A *parasitic array* consists of a single driven element connected to the feedline, and other elements which are not, called parasitic elements. It is usually another name for a Yagi-Uda antenna.

A *phased array* usually means an *electronically scanned array*; a driven array antenna in which each individual element is connected to the transmitter or receiver through a phase shifter controlled by a computer. The beam of radio waves can be steered electronically to point instantly in any direction over a wide angle. However the term "phased array" is sometimes used to mean an ordinary array antenna.

Description

Small antennas around one wavelength in size, such as quarter-wave monopoles and half-wave dipoles, don't have much directivity (gain); they are omnidirectional antennas which radiate radio waves over a wide angle. To create a high gain antenna, which radiates radio waves in a narrow beam, two general techniques can be used. One technique is to use large metal or dielectric surfaces such as parabolic reflectors, horns or lenses which change the direction of the radio waves by reflection or refraction, to focus the radio waves from a single low gain antenna into a beam.

A second technique is to use multiple antennas which are fed from the same transmitter or receiver; this is called an array antenna, or antenna array. If the currents are fed to the antennas with the proper phase, due to the phenomenon of interference the spherical waves from the individual antennas combine (superpose) in front of the array to create plane waves, a beam of radio waves traveling in a specific direction. In directions in which the waves from the individual antennas arrive in phase, the waves add together (constructive interference) to enhance the power radiated. In directions in which the individual waves arrive out of phase, with the peak of one wave coinciding with the valley of another, the waves cancel (destructive interference) reducing the power radiated in that direction. Similarly, when receiving, the currents from radio waves received from desired directions are in phase and when combined in the receiver reinforce each other, while currents from radio waves received from other directions are out of phase and when combined in the receiver cancel each other.

The radiation pattern of such an antenna consists of a strong beam in one direction, the main lobe, plus a series of weaker beams at different angles called sidelobes, usually representing residual radiation in unwanted directions. The larger the width of the antenna and the greater the number of component antenna elements, the narrower the main lobe, and the higher the gain which can be achieved, and the smaller the sidelobes will be.

Arrays in which the antenna elements are fed in phase are broadside arrays; the main lobe is emitted perpendicular to the plane of the elements.

The largest array antennas are radio interferometers used in the field of radio astronomy, in which multiple radio telescopes consisting of large parabolic antennas are linked together into an antenna array, to achieve higher resolution. Using the technique called aperture synthesis such an array can have the resolution of an antenna with a diameter equal to the distance between the antennas. In the technique called Very Long Baseline Interferometry (VLBI) dishes on separate continents have been linked, creating "array antennas" thousands of miles in size.

Antenna Analyzer

An antenna analyzer measuring SWR and complex impedance of a dummy load. MFJ Enterprises Inc. MFJ-269.

An antenna analyzer or in British aerial analyser (also known as a noise bridge, RX bridge, SWR analyzer, or RF analyzer) is a device used for measuring the input impedance of antenna systems in radio electronics applications.

Types of Analysers

Antenna Bridge

A bridge circuit has two legs which are frequency-dependent complex-valued impedances. One leg is a circuit in the analyzer with calibrated components whose combined impedance can be read on a scale. The other leg is the *unknown* – either an antenna or a reactive component.

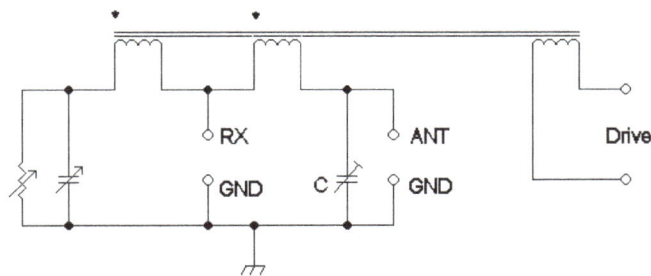

A typical antenna bridge, the trimmer capacitor (C) is adjusted to make the bridge balance when the variable capacitor on the left is half meshed. Hence the bridge will be able to detect if an antenna is either a capacitive or inductive load.

To measure impedance, the bridge is adjusted, so that the two legs have the same impedance. When the two impedances are the same, the bridge is balanced. Using this circuit it is possible to either measure the impedance of the antenna connected between ANT and GND, or it is possible to adjust an antenna, until it has the same impedance as the network on the left side of the diagram below. The bridge can be driven either with *white noise* or a simple carrier (connected to drive). In the case of white noise the amplitude of the exciting signal can be very low and a radio receiver used as the detector. In the case where a simple carrier is used then depending on the level either a diode detector or a receiver can be used. In both cases a null will indicate when the bridge is balanced.

Complex Voltage and Current Meters

A second type of antenna analyzer measures the complex voltage across and current into the antenna. It then uses mathematical methods to calculate complex impedance and display it in either on a calibrated meter or a digital display. Professional instruments of this type are usually called network analyzers. This type of analyzer does not require the operator to adjust any R and X knobs as with the bridge type analyzer. Many of these instruments have the ability to sweep the frequency and thus plot the antenna characteristics over a wide range. Doing this with a manually operated bridge would be time consuming, requiring one to change the frequency and adjust the knobs at each frequency for a null.

High and Low Power Methods

Many transmitters include an SWR meter in the output circuits which works by measuring the reflected wave from the antenna back to the transmitter, which is minimal when the antenna is matched. Reflected power from a badly tuned antenna can present an improper load at the transmitter which can damage it. A complex-impedance antenna analyzer typically requires only a few milliwatts of power be applied to the antenna. This avoids possible damage to the transmitter when the antenna is badly tuned. In addition, if the power is very low, the analyzer can be used outside of the frequency bands licensed to the transmitter operator and thus get data on the antenna performance over a range of frequencies.

Uses

In radio communications systems, including amateur radio, an antenna analyzer is a common tool used for troubleshooting antennas and feedlines as well as fine tuning their performance.

Antenna bridges have long been used in the broadcast industry to tune antennas. A bridge is available which measures complex impedance while the transmitter is operating, practically a necessity when tuning multi tower antenna systems. In more recent times the direct reading network analyzers have become more common.

Antenna Efficiency

Antenna efficiency, η, is a term associated with aperture antennas, e.g., parabolic dishes. Antenna efficiency is different from and contrasted with radiation efficiency, which applies to any antenna type.

Definition

Antenna efficiency is defined as the ratio of the aperture effective area, A_e to its actual physical area, A. It describes the percentage of the physical aperture area which actually captures radio frequency (RF) energy. Thus, the effective area of an aperture antenna is the surface area of a theoretically perfect aperture that would collect the same energy as the actual aperture with associated antenna efficiency

$$A_e = A\eta$$

Effective area is an important concept in the study of antennas, because the ratio of the gain of an antenna to its effective area can be shown to be a universal constant.

Other Definitions of Efficiency in Antennas

IEEE defines several other antenna parameters which include the word efficiency, such as

- aperture illumination efficiency for aperture-antennas.

- polarization efficiency; polarization mismatch factor.

These are closely related and easily misunderstood definitions. Care must be taken in engineering applications to ensure that the efficiency specifications are specific, clear, and unambiguous.

Radio Masts and Towers

Masts of the Rugby VLF transmitter in England

Radio masts and towers are, typically, tall structures designed to support antennas (also known as aerials) for telecommunications and broadcasting, including television. There are two main types: guyed and self-supporting structures. They are among the tallest man-made structures.

Masts are often named after the broadcasting organizations that originally built them or currently use them.

In the case of a mast radiator or radiating tower, the whole mast or tower is itself the transmitting antenna.

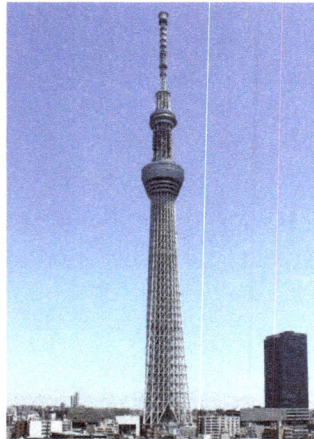

The Tokyo Skytree, the tallest freestanding tower in the world, in 2012

Mast or Tower

A radio mast base showing how virtually all lateral support is provided by the guy-wires

The terms "mast" and "tower" are often used interchangeably. However, in structural engineering terms, a tower is a self-supporting or cantilevered structure, while a mast is held up by stays or guys. Broadcast engineers in the UK use the same terminology. A mast is a ground-based or roof-top structure that supports antennas at a height where they can satisfactorily send or receive radio waves. Typical masts are of steel lattice or tubular steel construction. Masts themselves play no part in the transmission of mobile telecommunications. Masts (to use the civil engineering terminology) tend to be cheaper to build but require an extended area surrounding them to accommodate the guy wires. Towers are more commonly used in cities where land is in short supply.

There are a few borderline designs that are partly free-standing and partly guyed, called additionally guyed towers. For example:

- The Gerbrandy tower consists of a self-supporting tower with a guyed mast on top.

- The few remaining Blaw-Knox towers do the opposite: they have a guyed lower section surmounted by a freestanding part.

- Zendstation Smilde, a tall tower with a guyed mast on top (guys go to ground)

- Torre de Collserola, a guyed tower with a guyed mast on top (tower portion is not free-standing)

History

Experimental radio broadcasting began in 1905, and commercial radio broke through in the 1920s.

Until August 8, 1991, the Warsaw radio mast was the world's tallest supported structure on land; its collapse left the KVLY/KTHI-TV mast as the tallest. There are over 50 radio structures in the United States that are 600 m (1968.5 ft) or taller.

Materials

Typical 200 foot (61 m) triangular guyed lattice mast of an AM radio station in Mount Vernon, Washington, USA

Steel Lattice

The steel lattice is the most widespread form of construction. It provides great strength, low weight and wind resistance, and economy in the use of materials. Lattices of triangular cross-section are most common, and square lattices are also widely used. Guyed masts are often used; the supporting guy lines carry lateral forces such as wind loads, allowing the mast to be very narrow and simply constructed.

Russian TV tower, Penza

When built as a tower, the structure may be parallel-sided or taper over part or all of its height. When constructed of several sections which taper exponentially with height, in the manner of the Eiffel Tower, the tower is said to be an Eiffelized one. The Crystal Palace tower in London is an example.

Tubular Steel

Guyed masts are sometimes also constructed out of steel tubes. This construction type has the advantage that cables and other components can be protected from weather inside the tube and consequently the structure may look cleaner. These masts are mainly used for FM-/TV-broadcasting, but sometimes also as mast radiator. The big mast of Mühlacker transmitting station is a good example of this. A disadvantage of this mast type is that it is much more affected by winds than masts with open bodies. Several tubular guyed masts have collapsed. In the UK, the Emley Moor and Waltham TV stations masts collapsed in the 1960s. In Germany the Bielstein transmitter collapsed in 1985. Tubular masts were not built in all countries. In Germany, France, UK, Czech, Slovakia, Japan and the former Soviet Union, many tubular guyed masts were built, while there are nearly none in Poland or North America.

In several cities in Russia and Ukraine several tubular guyed masts with crossbars running from the mast structure to the guys were built in the 1960s. All these masts, which are designed as 30107 KM, are exclusively used for FM and TV transmission and, except for the mast in Vinnytsia, are between 150 and 200 metres tall. The crossbars of these masts are equipped with a gangway that holds smaller antennas, though their main purpose is oscillation damping.

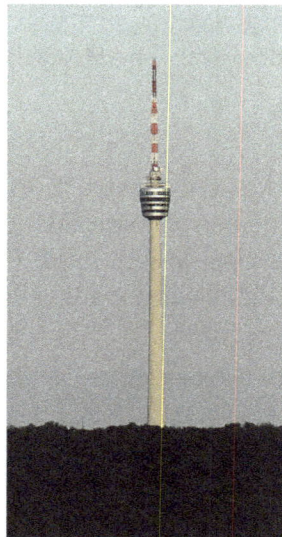

TV Tower in Stuttgart (Germany): the first reinforced-concrete TV tower

Reinforced Concrete

Reinforced concrete towers are relatively expensive to build but provide a high degree of mechanical rigidity in strong winds. This can be important when antennas with narrow beamwidths are used, such as those used for microwave point-to-point links, and when the structure is to be occupied by people.

In the 1950s, AT&T built numerous concrete towers, more resembling silos than towers, for its first transcontinental microwave route. Many are still in use today.

In Germany and the Netherlands most towers constructed for point-to-point microwave links are built of reinforced concrete, while in the UK most are lattice towers.

Tokyo Tower

Concrete towers can form prestigious landmarks, such as the CN Tower in Toronto. In addition to accommodating technical staff, these buildings may have public areas such as observation decks or restaurants.

The Stuttgart TV tower was the first tower in the world to be built in reinforced concrete. It was designed in 1956 by the local civil engineer Fritz Leonhardt.

Kamzík TV Tower, overlooking Bratislava (Slovakia).

Fiberglass and Other Composite Materials

Fiberglass poles are occasionally used for low-power non-directional beacons or medium-wave broadcast transmitters. Carbon fibre monopoles and towers have traditionally been too expensive but recent developments in the way the carbon fibre tow is spun have resulted in solutions that offer strengths similar or exceeding steel for a fraction of the weight - now allowing monopole and towers to be built in locations that were too expensive or difficult to access with the heavy lifting equipment that is needed for steel structure.

Wood

There are fewer wooden towers now than in the past. Many were built in the UK during World War II because of a shortage of steel. In Germany before World War II wooden towers were used at nearly all medium-wave transmission sites, but all of these towers have since been demolished, except for the Gliwice Radio Tower.

Ferryside television relay station is an example of a TV relay transmitter using a wooden pole.

Other types of Antenna Supports and Structures

Poles

Shorter masts may consist of a self-supporting or guyed wooden pole, similar to a telegraph pole. Sometimes self-supporting tubular galvanized steel poles are used: these may be termed monopoles.

Buildings

In some cases, it is possible to install transmitting antennas on the roofs of tall buildings. In North America, for instance, there are transmitting antennas on the Empire State Building, the Willis Tower, 4 Times Square, and One World Trade Center. The North Tower (1WTC) of the original World Trade Center also had a 360-foot (110m) telecommunications antenna atop its roof, constructed in 1978-1979, and began transmission in 1980. When the buildings collapsed, several local TV and radio stations were knocked off the air until backup transmitters could be put into service. Such facilities also exist in Europe, particularly for portable radio services and low-power FM radio stations. In London, the BBC erected in 1936 a mast for broadcasting early television on one of the towers of a Victorian building, the Alexandra Palace. It is still in use.

Disguised Cell-sites

Completed in December 2009 at Epiphany Lutheran Church in Lake Worth, Florida, this 100' tall cross conceals equipment for T-Mobile

Many people view bare cellphone towers as ugly and an intrusion into their neighbourhoods. Even though people increasingly depend upon cellular communications, they are opposed to the bare towers spoiling otherwise scenic views. Many companies offer to 'hide' cellphone towers in, or as, trees, church towers, flag poles, water tanks and other features. There are many providers that offer these services as part of the normal tower installation and maintenance service. These are generally called "stealth towers" or "stealth installations", or simply concealed cell sites.

The level of detail and realism achieved by disguised cellphone towers is remarkably high; for example, such towers disguised as trees are nearly indistinguishable from the real thing, even for local wildlife (who additionally benefit from the artificial flora). Such towers can be placed unobtrusively in national parks and other such protected places, such as towers disguised as cacti in Coronado National Forest.

Even when disguised, however, such towers can create controversy; a tower doubling as a flagpole attracted controversy in 2004 in relation to the U.S. Presidential campaign of that year, and highlighted the sentiment that such disguises serve more to allow the installation of such towers in subterfuge away from public scrutiny rather than to serve towards the beautification of the landscape.

Disguised cell sites sometimes can be introduced into environments that require a low-impact visual outcome, by being made to look like trees, chimneys or other common structures.

Mast Radiators

A mast radiator is a radio tower or mast in which the whole structure works as an antenna. It is used frequently as a transmitting antenna for long or medium wave broadcasting.

Structurally, the only difference is that a mast radiator may be supported on an insulator at its base. In the case of a tower, there will be one insulator supporting each leg.

Telescopic, Pump-up and Tiltover Towers

A special form of the radio tower is the *telescopic mast*. These can be erected very quickly. Telescopic masts are used predominantly in setting up temporary radio links for reporting on major news events, and for temporary communications in emergencies. They are also used in tactical military networks. They can save money by needing to withstand high winds only when raised, and as such are widely used in amateur radio.

Telescopic masts consist of two or more concentric sections and come in two principal types:

- Pump-up masts are often used on vehicles, and are raised to their full height pneumatically or hydraulically. They are usually only strong enough to support fairly small antennas.

- Telescopic lattice masts are raised by means of a winch, which may be powered by hand or an electric motor. These tend to cater for greater heights and loads than the pump-up type. When retracted, the whole assembly can sometimes be lowered to a horizontal position by means of a second tiltover winch. This enables antennas to be fitted and adjusted at ground level before winching the mast up.

Balloons and kites

A tethered balloon or a kite can serve as a temporary support. It can carry an antenna or a wire (for VLF, LW or MW) up to an appropriate height. Such an arrangement is used occasionally by military agencies or radio amateurs. The American broadcasters TV Martí broadcast a television program to Cuba by means of such a balloon.

Drones

There has recently (2013) been interest in using unmanned aerial vehicles (drones) for telecom purposes. It is not clear what advantages a drone would have over a balloon.

Other Special Structures

For two VLF transmitters wire antennas spun across deep valleys are used. The wires are supported by small masts or towers or rock anchors. See List of spans: Antenna spans across valleys. The same technique was also used at Criggion radio station.

For ELF transmitters ground dipole antennas are used. Such structures require no tall masts. They consist of two electrodes buried deep in the ground at least a few dozen kilometres apart. From the transmitter building to the electrodes, overhead feeder lines run. These lines look like power lines of the 10 kV level, and are installed on similar pylons.

Design Features

Economic and Aesthetic Considerations

A radio amateur's do it yourself steel-lattice tower

Felsenegg-Girstel TV-tower

Uetliberg TV-tower

Communications tower, camouflaged as a slim tree

- The cost of a mast or tower is roughly proportional to the square of its height.

- A guyed mast is cheaper to build than a self-supporting tower of equal height.

- A guyed mast needs additional land to accommodate the guys, and is thus best suited to rural locations where land is relatively cheap. An unguyed tower will fit into a much smaller plot.

- A steel lattice tower is cheaper to build than a concrete tower of equal height.

- Two small towers may be less intrusive, visually, than one big one, especially if they look identical.

- Towers look less ugly if they and the antennas mounted on them appear symmetrical.

- Concrete towers can be built with aesthetic design - and they are, especially in Continental Europe. They are sometimes built in prominent places and include observation decks or restaurants.

Masts for HF/shortwave Antennas

For transmissions in the shortwave range, there is little to be gained by raising the antenna more than a few wavelengths above ground level. Shortwave transmitters rarely use masts taller than about 100 metres.

Access for Riggers

Because masts, towers and the antennas mounted on them require maintenance, access to the whole of the structure is necessary. Small structures are typically accessed with a ladder. Larger structures, which tend to require more frequent maintenance, may have stairs and sometimes a lift, also called a service elevator.

Aircraft Warning Features

Tall structures in excess of certain legislated heights are often equipped with aircraft warning lamps, usually red, to warn pilots of the structure's existence. In the past, ruggedized and under-run filament lamps were used to maximize the bulb life. Alternatively, neon lamps were used. Nowadays such lamps tend to use LED arrays.

Height requirements vary across states and countries, and may include additional rules such as requiring a white flashing strobe in the daytime and pulsating red fixtures at night. Structures over a certain height may also be required to be painted with contrasting color schemes such as white and orange or white and red to make them more visible against the sky.

Light Pollution and Nuisance Lighting

In some countries where light pollution is a concern, tower heights may be restricted so as to reduce or eliminate the need for aircraft warning lights. For example, in the United States the 1996 Telecommunications Act allows local jurisdictions to set maximum heights for towers, such as limiting tower height to below 200 feet and therefore not requiring aircraft illumination under U.S. Federal Communications Commission (FCC) rules. The limit is more commonly set to 190 or 180 feet to allow for masts extending above the tower.

Wind-induced Oscillations

One problem with radio masts is the danger of wind-induced oscillations. This is particularly a concern with steel tube construction. One can reduce this by building cylindrical shock-mounts into the construction. One finds such shock-mounts, which look like cylinders thicker than the mast, for example, at the radio masts of DHO38 in Saterland. There are also constructions, which consist of a free-standing tower (usually from reinforced concrete), onto which a guyed radio mast is installed. The best known such construction is the Gerbrandy Tower in Lopik (the Netherlands). Further towers of this building method can be found near Smilde (the Netherlands) and Fernseh-turm, Waldenburg, Baden-Württemberg, Germany).

Hazard to Birds

Radio, television and cell towers have been documented to pose a hazard to birds. Reports have been issued documenting known bird fatalities and calling for research to find ways to minimize the hazard that communications towers can pose to birds. There have also been instances of rare birds nesting in cell towers and thereby preventing repair work due to legislation intended to protect them.

Catastrophic Collapses

Law

Since June 2010, Telecom operators in the USA can erect new telecom masts or towers as the government has lifted the moratorium, which was earlier placed on the issuance of permits for the construction of telecommunication towers.

Antenna Measurement

Antenna measurement techniques refers to the testing of antennas to ensure that the antenna meets specifications or simply to characterize it. Typical parameters of antennas are gain, radiation pattern, beamwidth, polarization, and impedance.

The antenna pattern is the response of the antenna to a plane wave incident from a given direction or the relative power density of the wave transmitted by the antenna in a given direction. For a reciprocal antenna, these two patterns are identical. A multitude of antenna pattern measurement techniques have been developed. The first technique developed was the far-field range, where the antenna under test (AUT) is placed in the far-field of a range antenna. Due to the size required to create a far-field range for large antennas, near-field techniques were developed, which allow the measurement of the field on a surface close to the antenna (typically 3 to 10 times its wavelength). This measurement is then predicted to be the same at infinity. A third common method is the compact range, which uses a reflector to create a field near the AUT that looks approximately like a plane-wave.

Far-field Range (FF)

The far-field range was the original antenna measurement technique, and consists of placing the AUT a long distance away from the instrumentation antenna. Generally, the far-field distance or Fraunhofer distance, d, is considered to be

$$d = \frac{2D^2}{\lambda},$$

where D is the maximum dimension of the antenna and λ is the wavelength of the radio wave. Separating the AUT and the instrumentation antenna by this distance reduces the phase variation across the AUT enough to obtain a reasonably good antenna pattern.

IEEE suggests the use of their antenna measurement standard, document number IEEE-Std-149-1979 for far-field ranges and measurement set-up for various techniques including ground-bounce type ranges.

Near-field Range (NF)

Planar Near-field Range

Planar near-field measurements are conducted by scanning a small probe antenna over a planar surface. These measurements are then transformed to the far-field by use of a Fourier transform, or more specifically by applying a method known as stationary phase to the Laplace transform . Three basic types of planar scans exist in near field measurements.

Rectangular Planar Scanning

The probe moves in the Cartesian coordinate system and its linear movement creates a regular rectangular sampling grid with a maximum near-field sample spacing of $\Delta x = \Delta y = \lambda / 2$.

Polar Planar Scanning

More complicated solution to the rectangular scanning method is the plane polar scanning method.

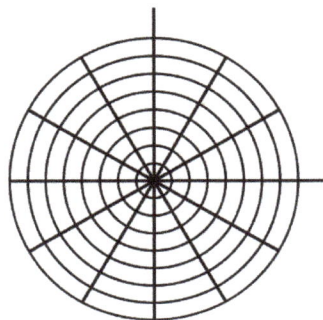

Bi-polar Planar Scanning

The bi-polar technique is very similar to the plane polar configuration.

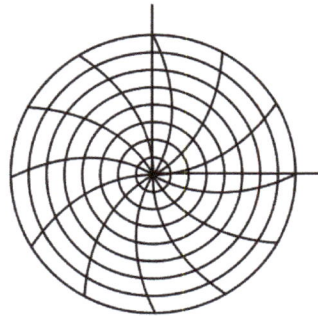

Cylindrical Near-field Range

Cylindrical near-field ranges measure the electric field on a cylindrical surface close to the AUT. Cylindrical harmonics are used transform these measurements to the far-field.

Spherical Near-field Range

Spherical near-field ranges measure the electric field on a spherical surface close to the AUT. Spherical harmonics are used transform these measurements to the far-field

Free-space Ranges

The formula for electromagnetic radiation dispersion and information is:

$$D^2 = \frac{P}{S} \propto 3dB$$

Where D=Distance, P=Power, and S=Speed

What this means is that double the communication distance requires four times the power. It also means double power allows double communication speed (bit rate). Double power is approx. 3dB (10 log(2) to be exact) increase. Of course in the real world there are all sorts of other phenomena which enter in, such as Fresnel canceling, path loss, background noise, etc.

Compact Range

A Compact Antenna Test Range (CATR) is a facility which is used to provide convenient testing of antenna systems at frequencies where obtaining far-field spacing to the AUT would be infeasible using traditional free space methods. It was invented by Richard C. Johnson at the Georgia Tech Research Institute. The CATR uses a source antenna which radiates a spherical wavefront and one or more secondary reflectors to collimate the radiated spherical wavefront into a planar wavefront within the desired test zone. One typical embodiment uses a horn feed antenna and a parabolic reflector to accomplish this.

The CATR is used for microwave and millimeter wave frequencies where the $2 D^2/\lambda$ far-field distance is large, such as with high-gain reflector antennas. The size of the range that is required can be much less than the size required for a full-size far-field anechoic chamber, although the cost of fabrication of the specially-designed CATR reflector can be expensive due to the need to ensure precision of the reflecting surface (typically less than $\lambda/100$ RMS surface accuracy) and to specially treat the edge of the reflector to avoid diffracted waves which can interfere with the desired beam pattern.

Elevated Range

A means of reducing reflection from waves bouncing off the ground.

Slant Range

A means of eliminating symmetrical wave reflection.

Antenna Parameters

Except for polarization, the SWR is the most easily measured of the parameters above. Impedance can be measured with specialized equipment, as it relates to the complex SWR. Measuring radiation pattern requires a sophisticated setup including significant clear space (enough to put the sensor into the antenna's far field, or an anechoic chamber designed for antenna measurements), careful study of experiment geometry, and specialized measurement equipment that rotates the antenna during the measurements.

Radiation Pattern

The radiation pattern is a graphical depiction of the relative field strength transmitted from or received by the antenna, and shows sidelobes and backlobes. As antennas radiate in space often several curves are necessary to describe the antenna. If the radiation of the antenna is symmetrical about an axis (as is the case in dipole, helical and some parabolic antennas) a unique graph is sufficient.

Each antenna supplier/user has different standards as well as plotting formats. Each format has its own advantages and disadvantages. Radiation pattern of an antenna can be defined as the locus of all points where the emitted power per unit surface is the same. The radiated power per unit surface is proportional to the squared electrical field of the electromagnetic wave. The radiation pattern is the locus of points with the same electrical field. In this representation, the reference is usually the best angle of emission. It is also possible to depict the directive gain of the antenna as a function of the direction. Often the gain is given in decibels.

The graphs can be drawn using cartesian (rectangular) coordinates or a polar plot. This last one is useful to measure the beamwidth, which is, by convention, the angle at the -3dB points around the max gain. The shape of curves can be very different in cartesian or polar coordinates and with the choice of the limits of the logarithmic scale. The four drawings below are the radiation patterns of a same half-wave antenna.

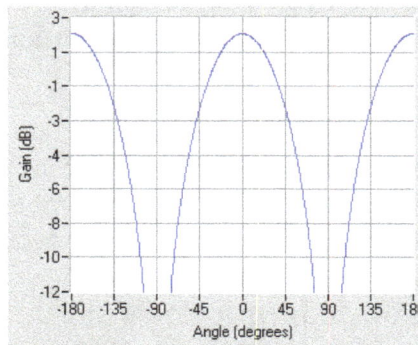

Gain of a half-wave dipole. Cartesian representation.

Efficiency

Efficiency is the ratio of power actually radiated by an antenna to the electrical power it receives from a transmitter. A dummy load may have an SWR of 1:1 but an efficiency of 0, as it absorbs all the incident power, producing heat but radiating no RF energy; SWR is not a measure of an antenna's efficiency. Radiation in an antenna is caused by radiation resistance which cannot be directly measured but is a component of the total resistance which includes the loss resistance. Loss resistance results in heat generation rather than radiation, thus reducing efficiency. Mathematically, efficiency is equal to the radiation resistance divided by total resistance (real part) of the feed-point impedance. Efficiency is defined as the ratio of the power that is radiated to the total power used by the antenna; Total power = power radiated + power loss.

$$\tilde{n} = \frac{P_r}{P_r + P_l}$$

Bandwidth

IEEE defines bandwidth as "The range of frequencies within which the performance of the antenna, with respect to some characteristic, conforms to a specified standard." In other words, bandwidth depends on the overall effectiveness of the antenna through a range of frequencies, so all of these parameters must be understood to fully characterize the bandwidth capabilities of an antenna. This definition may serve as a practical definition, however, in practice, bandwidth is typically determined by measuring a characteristic such as SWR or radiated power over the frequency range of interest. For example, the SWR bandwidth is typically determined by measuring the frequency range where the SWR is less than 2:1. Another frequently used value for determining bandwidth for resonant antennas is the -3dB Return Loss value.

Directivity

Antenna directivity is the ratio of maximum radiation intensity (power per unit surface) radiated by the antenna in the maximum direction divided by the intensity radiated by a hypothetical isotropic antenna radiating the same total power as that antenna. For example, a hypothetical antenna which had a radiated pattern of a hemisphere (1/2 sphere) would have a directivity of 2. Directivity is a dimensionless ratio and may be expressed numerically or in decibels (dB). Directivity is identical to the peak value of the directive gain; these values are specified without respect to antenna efficiency thus differing from the power gain (or simply "gain") whose value *is* reduced by an antenna's efficiency.

Gain

Gain as a parameter measures the directionality of a given antenna. An antenna with a low gain emits radiation in all directions equally, whereas a high-gain antenna will preferentially radiate in particular directions. Specifically, the **Gain** or **Power gain** of an antenna is defined as the ratio of the intensity (power per unit surface) radiated by the antenna in a given direction at an arbitrary distance divided by the intensity radiated at the same distance by an hypothetical isotropic antenna:

$$G = \frac{\left(\dfrac{P}{S}\right)_{ant}}{\left(\dfrac{P}{S}\right)_{iso}}$$

We write "hypothetical" because a perfect isotropic antenna cannot be constructed. Gain is a dimensionless number (without units).

The gain of an antenna is a passive phenomenon - power is not added by the antenna, but simply redistributed to provide more radiated power in a certain direction than would be transmitted by an isotropic antenna. If an antenna has a greater than one gain in some directions, it must have a less than one gain in other directions since energy is conserved by the antenna. An antenna designer must take into account the application for the antenna when determining the gain. High-gain antennas have the advantage of longer range and better signal quality, but must be aimed carefully in a particular direction. Low-gain antennas have shorter range, but the orientation of the antenna is inconsequential. For example, a dish antenna on a spacecraft is a high-gain device (must be pointed at the planet to be effective), while a typical WiFi antenna in a laptop computer is low-gain (as long as the base station is within range, the antenna can be in an any orientation in space).

As an example, consider an antenna that radiates an electromagnetic wave whose electrical field has an amplitude E_θ at a distance r. This amplitude is given by:

$$E_\theta = \frac{AI}{r}$$

where:

I is the current fed to the antenna and

A is a constant characteristic of each antenna.

For a large distance r. The radiated wave can be considered locally as a plane wave. The intensity of an electromagnetic plane wave is:

$$\frac{P}{S} = \frac{c\varepsilon_\circ}{2} E_\theta^{\ 2} = \frac{1}{2} \frac{E_\theta^{\ 2}}{Z_\circ}$$

where $Z_\circ = \sqrt{\dfrac{\mu_\circ}{\varepsilon_\circ}} = 376.7303134612\Omega$ is a universal constant called vacuum impedance. and

$$\left(\frac{P}{S}\right)_{ant} = \frac{1}{2Z_\circ} \frac{A^2 I^2}{r^2}$$

If the resistive part of the series impedance of the antenna is R_s, the power fed to the antenna is $\frac{1}{2} R_s I^2$. The intensity of an isotropic antenna is the power so fed divided by the surface of the sphere of radius r:

$$\left(\frac{P}{S}\right)_{iso} = \frac{\frac{1}{2} R_s I^2}{4\pi r^2}$$

The directive gain is:

$$G = \frac{\dfrac{1}{2Z_\circ} \dfrac{A^2 I^2}{r^2}}{\dfrac{\frac{1}{2} R_s I^2}{4\pi r^2}} = \frac{A^2}{30 R_s}$$

For the commonly utilized half-wave dipole, the particular formulation works out to the following, including its decibel equivalency, expressed as dBi (decibels referenced to **i**sotropic radiator):

$$R_{\frac{\lambda}{2}} = 60 Cin(2\pi) = 60\left[\ln(2\pi\gamma) - Ci(2\pi)\right] = 120\int_0^{\frac{\pi}{2}} \frac{\cos^2\left(\frac{\pi}{2}\cos\theta\right)}{\sin\theta}\, d\theta,$$

$$= 15\left[2\pi^2 - \frac{1}{3}\pi^4 + \frac{4}{135}\pi^6 - \frac{1}{630}\pi^8 + \frac{4}{70875}\pi^{10} \ldots - (-1)^n \frac{(2\pi)^{2n}}{n(2n)!}\right],$$

$$= 73.1296\ldots\Omega;$$

(In most cases 73.13, is adequate)

$$G_{\frac{\lambda}{2}} = \frac{60^2}{30 R_{\frac{\lambda}{2}}} = \frac{3600}{30 R_{\frac{\lambda}{2}}} = \frac{120}{R_{\frac{\lambda}{2}}} = \frac{1}{\int_0^{\frac{\pi}{2}} \frac{\cos^2\left(\frac{\pi}{2}\cos\theta\right)}{\sin\theta}\, d\theta},$$

$$\approx \frac{120}{73.1296} \approx 1.6409224 \approx 2.15088 \text{ dBi};$$

(Likewise, 1.64 and 2.15 dBi are usually the cited values)

Sometimes, the half-wave dipole is taken as a reference instead of the isotropic radiator. The gain is then given in dBd (decibels over dipole):

0 dBd = 2.15 dBi

Physical Background

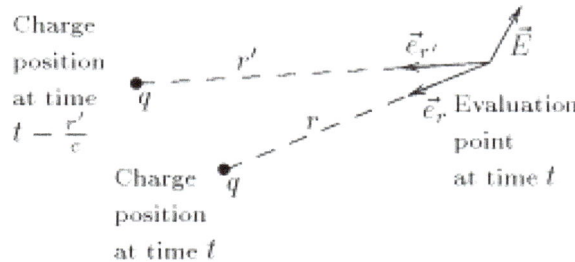

The measured electrical field was radiated $\dfrac{r'}{c}$ seconds earlier.

The electrical field created by an electric charge q is

$$\vec{E} = \frac{-q}{4\pi\varepsilon_\circ}\left[\frac{\vec{e}_{r'}}{r'^2} + \frac{r'}{c}\frac{d}{dt}\left(\frac{\vec{e}_{r'}}{r'^2}\right) + \frac{1}{c^2}\frac{d^2}{dt^2}\left(\vec{e}_{r'}\right)\right]$$

where:

c is the speed of light in vacuum.

$\varepsilon_\circ$ is the permittivity of free space.

r' is the distance from the observation point (the place where $\vec{E}$ is evaluated) to the point where the charge *was* $\dfrac{r'}{c}$ seconds *before* the time when the measure is done.

$\vec{e}_{r'}$ is the unit vector directed from the observation point (the place where $\vec{E}$ is evaluated) to the point where the charge *was* $\dfrac{r'}{c}$ seconds *before* the time when the measure is done.

The "prime" in this formula appears because the electromagnetic signal travels at the speed of light. Signals are observed as coming from the point where they were emitted and not from the point where the emitter is at the time of observation. The stars that we see in the sky are no longer where we see them. We will see their current position years in the future; some of the stars that we see today no longer exist.

The first term in the formula is just the electrostatic field with retarded time.

The second term is *as though nature were trying to allow for the fact that the effect is retarded* (Feynman).

The third term is the only term that accounts for the far field of antennas.

The two first terms are proportional to $\dfrac{1}{r^2}$. Only the third is proportional to $\dfrac{1}{r}$.

Near the antenna, all the terms are important. However, if the distance is large enough, the first two terms become negligible and only the third remains:

$$\vec{E} = \frac{-q}{4\pi\varepsilon c_\circ^2}\frac{d^2}{dt^2}(\vec{e}_{r'}) = -q10^{-7}\frac{d^2}{dt^2}(\vec{e}_{r'})$$

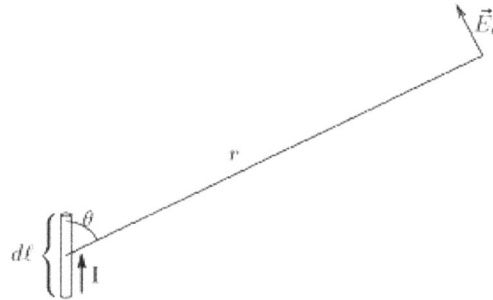

Electrical field radiated by an element of current. The element of current, the electrical field vector $\vec{E}_\theta$ and r are on the same plane.

If the charge q is in sinusoidal motion with amplitude $\ell_\circ$ and pulsation ω the power radiated by the charge is:

$$P = \frac{q^2\omega^4\ell_\circ^2}{12\pi\varepsilon_\circ c^3} \text{ watts.}$$

Note that the radiated power is proportional to the fourth power of the frequency. It is far easier to radiate at high frequencies than at low frequencies. If the motion of charges is due to currents, it can be shown that the (small) electrical field radiated by a small length $d\ell$ of a conductor carrying a time varying current I is

$$dE_\theta(t+\frac{r}{c}) = \frac{-d\ell\sin\theta}{4\pi\varepsilon_\circ c^2 r}\frac{dI}{dt}$$

The left side of this equation is the electrical field of the electromagnetic wave radiated by a small length of conductor. The index θ reminds that the field is perpendicular to the line to the source. The $t+\dfrac{r}{c}$ reminds that this is the field observed $\dfrac{r}{c}$ seconds after the evaluation on the current derivative. The angle θ is the angle between the direction of the current and the direction to the point where the field is measured.

The electrical field and the radiated power are maximal in the plane perpendicular to the current element. They are zero in the direction of the current.

Only time-varying currents radiate electromagnetic power.

If the current is sinusoidal, it can be written in complex form, in the same way used for impedances. Only the real part is physically meaningful:

$$I = I_\circ e^{j\omega t}$$

where:

$I_\circ$ is the amplitude of the current.

$\omega = 2\pi f$ is the angular frequency.

$j = \sqrt{-1}$

The (small) electric field of the electromagnetic wave radiated by an element of current is:

$$dE_\theta\left(t+\frac{r}{c}\right) = \frac{-d\ell j\omega}{4\pi\varepsilon_\circ c^2}\frac{\sin\theta}{r}e^{j\omega t}$$

And for the time t :

$$dE_\theta(t) = \frac{-d\ell j\omega}{4\pi\varepsilon_\circ c^2}\frac{\sin\theta}{r}e^{j\left(\omega t-\frac{\omega}{c}r\right)}$$

The electric field of the electromagnetic wave radiated by an antenna formed by wires is the sum of all the electric fields radiated by all the small elements of current. This addition is complicated by the fact that the direction and phase of each of the electric fields are, in general, different.

Calculation of Antenna Parameters in Reception

The gain in any given direction and the impedance at a given frequency are the same when the antenna is used in transmission or in reception.

The electric field of an electromagnetic wave induces a small voltage in each small segment in all electric conductors. The induced voltage depends on the electrical field and the conductor length. The voltage depends also on the relative orientation of the segment and the electrical field.

Each small voltage induces a current and these currents circulate through a small part of the antenna impedance. The result of all those currents and tensions is far from immediate. However, using the reciprocity theorem, it is possible to prove that the Thévenin equivalent circuit of a receiving antenna is:

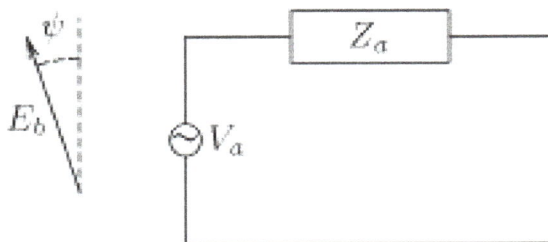

$$V_a = \frac{\sqrt{R_a G_a}\,\lambda\cos\psi}{2\sqrt{\pi Z_\circ}}E_b$$

where:

V_a is the Thévenin equivalent circuit tension.

Z_a is the Thévenin equivalent circuit impedance and is the same as the antenna impedance.

R_a is the series resistive part of the antenna impedance Z_a.

G_a is the directive gain of the antenna (the same as in emission) in the direction of arrival
of electromagnetic waves.

λ is the wavelength.

E_b is the magnitude of the electrical field of the incoming electromagnetic wave.

ψ is the angle of misalignment of the electrical field of the incoming wave with the antenna. For a dipole antenna, the maximum induced voltage is obtained when the electrical
field is parallel to the dipole. If this is not the case and they are misaligned by an angle ψ,
the induced voltage will be multiplied by $\cos\psi$.

$Z_\circ = \sqrt{\dfrac{\mu_\circ}{\varepsilon_\circ}} = 376.730313461\,\Omega$ is a universal constant called vacuum impedance or impedance of free space.

The equivalent circuit and the formula at right are valid for any type of antenna. It can be as well a
dipole antenna, a loop antenna, a parabolic antenna, or an antenna array.

From this formula, it is easy to prove the following definitions:

$$\text{Antenna effective length} = \frac{\sqrt{R_a G_a}\,\lambda\cos\psi}{\sqrt{\pi Z_\circ}}$$

is the length which, multiplied by the electrical field of the received wave, give the voltage of the
Thévenin equivalent antenna circuit.

$$\text{Maximum available power} = \frac{G_a \lambda^2}{4\pi Z_\circ} E_b^2$$

is the maximum power that an antenna can extract from the incoming electromagnetic wave.

$$\text{Cross section or effective capture surface} = \frac{G_a}{4\pi}\lambda^2$$

is the surface which multiplied by the power per unit surface of the incoming wave, gives the maximum available power.

The maximum power that an antenna can extract from the electromagnetic field depends only on the
gain of the antenna and the squared wavelength λ. It does not depend on the antenna dimensions.

Using the equivalent circuit, it can be shown that the maximum power is absorbed by the antenna when it is terminated with a load matched to the antenna input impedance. This also implies that under matched conditions, the amount of power re-radiated by the receiving antenna is equal to that absorbed.

References

- Lonngren, Karl Erik; Savov, Sava V.; Jost, Randy J. (2007). Fundamentals of Electomagnetics With Matlab, 2nd Ed. SciTech Publishing. p. 451. ISBN 1891121588

- Bevelaqua, Peter J. "Types of Antennas". Antenna Theory. Antenna-theory.com Peter Bevelaqua's private website. Retrieved June 28, 2015

- Stutzman, Warren L.; Thiele, Gary A. (2012). Antenna Theory and Design, 3rd Ed. John Wiley & Sons. pp. 560–564. ISBN 0470576642

- Galindo, V. (1964). "Design of dual-reflector antennas with arbitrary phase and amplitude distributions". Antennas and Propagation, IEEE Transactions on. IEEE. 12 (4): 403–408. doi:10.1109/TAP.1964.1138236

- "Dipole Antenna / Aerial tutorial". Resources. Radio-Electronics.com, Adrio Communications, Ltd. 2011. Retrieved April 29, 2013

- Silver, H. Ward, ed. (2011). ARRL Antenna Book. Newington, Connecticut: American Radio Relay League. p. 3-2. ISBN 978-0-87259-694-8

- "BLIND FLYING ON THE BEAM: AERONAUTICAL COMMUNICATION, NAVIGATION AND SURVEILLANCE: ITS ORIGINS AND THE POLITICS OF TECHNOLOGY" (PDF). Journal of Air Transportation. 2003

- Poole, Ian (2016). "What is MIMO? Multiple Input Multiple Output Tutorial". Antennas and propagation. Radio-electronics.com (Adrio Communications. Retrieved February 23, 2017

- Stutzman, Warren L.; Thiele, Gary A. (2012). Antenna Theory and Design. John Wiley and Sons. pp. 74–75. ISBN 0470576642

- Ray, Bill (17 April 2013). "Angry Birds fire back: Vulture cousins menace UK city's mobiles". The Register. Retrieved 20 May 2013

- R. K. Mongia, and P. Bhartia, "Dielectric Resonator Antennas – A Review and General Design Relations for Resonant Frequency and Bandwidth", International Journal of Microwave and Millimeter-Wave Computer-Aided Engineering, 1994, 4, (3), pp 230–247

- Straw, R. Dean, Ed. (2000). The ARRL Antenna Book, 19th Ed. USA: American Radio Relay League. p. 19.15. ISBN 0-87259-817-9

Permissions

Index

www.ingramcontent.com/pod-product-compliance
Lightning Source LLC
Chambersburg PA
CBHW061243190326
41458CB00011B/3567